T0215796

Environmental Chemistry

Environmental Chemistry

Microscale Laboratory Experiments

Jorge G. Ibanez
Margarita Hernandez-Esparza
Carmen Doria-Serrano
Arturo Fregoso-Infante
Mexican Microscale Chemistry Center
Dept. of Chemical and Engineering Sciences
Universidad Iberoamericana—Mexico City

Mono Mohan Singh
National Microscale Chemistry Center
Merrimack College—North Andover, MA

 Springer

Library of Congress Control Number: 2007920429

ISBN 978-0-387-49492-0
ISBN 978-0-387-49492-8 (eBook)

Printed on acid-free paper.

9 8 7 6 5 4 3 2 1

springer.com

Foreword

Environmental Chemistry: Fundamentals and Microscale Experiments

When I was about eight years old, the beautiful wetland that stretched out beside my home was destroyed and was replaced by a large office park. My father, upon seeing how upset I was, said to me, "If you care about something, you care enough to learn about it." This event perhaps more than anything else caused me to dedicate my life to sustaining our environment, our world, through the power of chemistry. And perhaps, more than anything else, this is also the lesson of these excellent textbooks, Environmental Chemistry: Fundamentals and Microscale Experiments. It is the knowledge and perspective contained in these textbooks that allows all of us as scientists to understand the way our environment functions on a molecular level and how to identify potential threats to human health and the environment that need to be addressed. But they do more than that. They also give us the fundamental basis for ensuring that those problems never arise by using the Principles of Green Chemistry that emphasize avoiding hazards through innovations in chemistry.

For much of the history of the environmental movement, the scientific community has sought to identify and quantify environmental problems. If we were able to identify the problems, we then sought ways to clean them up sometimes elegantly, sometimes expensively, sometimes both. Because of the knowledge and fundamentals presented in these textbooks we are able to build on those historical approaches and incorporate the principles of sustainable design into the chemical products and processes of tomorrow. The authors deserve tremendous credit for their extraordinary efforts that have resulted in these comprehensive and much needed volumes. These books will provide excellent resources for those aspiring scientists to understand that the way to protect the Earth is through intimately and rigorously understanding the Earth.

Paul T. Anastas
Yale University
December, 2006

Preface

Modern science is not straightforward. Intricate relationships exist among the different disciplines involved in the understanding of virtually every scientific issue and phenomenon. The days of the Renaissance, when a single person could master a large portion of the knowledge then available, are long gone. This is not due to a lack of individual capacity, but rather to the explosion of knowledge, characteristic of our times.

Environmental Science—and more specifically, Environmental Chemistry—finds itself completely immersed in such a scenario. In this regard, a book written by several authors having complementary backgrounds and interests appeared to be an appropriate project to pursue. On the other hand, multiauthored undergraduate textbooks run the risk of lacking smoothness and continuity in the presentation of ideas and concepts. The present project involved many meetings and cross-checking among the five authors. This is why we perceive this finished task as valuable, and we hope that the reader finds a flowing progression and fair treatment of the various subjects.

The book is written with sophomore or junior college students in mind (i.e., undergraduate students in their second or third year). However, issues are often presented in such a way that General Chemistry students—and even graduate students—can find subjects of interest applicable to their level. The book consists of a theoretical section (12 chapters) and a companion book with an experimental section (24 experiments) in two separate volumes. A brief description now follows (initials of the main authors of each chapter and experiment appear in parentheses).

The beginning of the theoretical section comprises a general introduction to Environmental Chemistry (Chapter 1, MH), and a summary of the main background concepts that a student of Environmental Chemistry ought to know (Chapter 2, JI; Chapter 3, JI, AF, MS). We assume that the students have the minimum background in Organic Chemistry and in Biochemistry necessary for Environmental Chemistry. Subsequent chapters discuss the composition and characteristics of the natural chemical processes that occur in the atmosphere (Chapter 4, AF), the lithosphere (Chapter 5, JI), and the hydrosphere (Chapter 6, MH and JI). This discussion concludes by examining natural biochemical processes and introducing the organisms in the biosphere (Chapter 7, CD). Chapters that follow then analyze the effects of many pollutants (Chapter 8, JI, CD and MS; Chapter 9, CD), their treatment (Chapter 10, JI and CD; Chapter 11, CD), and the minimization and prevention of pollution, emphasizing Green Chemistry (Chapter 12, CD). Each chapter also contains a list of educational experiments in the literature related to its subject and a list of other useful references.

The experiments are rather varied, ranging from the characterization of aqueous media to pollutant-treatment schemes. For increased safety, savings, and environmental awareness, as well as for reduced costs, wastes, and environmental damage, we present our experiments at the *microscale* level (sometimes also called *small-scale*). Such experiments typically use microliters or micromoles of at least one of the reagents. The main authors of each experiment are as follows: Experiments 1–4, MH; 5, 6, JI; 7, AF and JI; 8, JI; 9, MH; 10, AF;11,12, JI; 13,14, JI and MS; 15, JI; 16, JI and AF; 17, JI and MH; 18–20, JI;

21, JI (from an experiment by Viktor Obendrauf); 22–24, CD. The answers to selected problems from the various experiments are given in the book's website at www.springer.com. Many open-ended projects are suggested in the *additional related projects* section of each experiment.

The possibilities for accidents or personal injury while performing these experiments are fairly small. However, owing to the incalculable number of variables involved when many individuals perform experiments in separate places with different materials and reagents, we cannot accept any responsibility in such unlikely events. In the same vein, we cannot accept responsibility for any possible consequences when performing the *additional related projects* described above.

The books contain a total of 240 questions, problems, and examples; of these, over 100 are solved in the text. They also contain more than 150 figures, 70 tables, and 1300 references to the literature (almost 50% of these references are related to educational environmental activities and experiments). Lastly, 80 *additional related projects* are suggested in the experimental section.

Further technical notes are in order:

(a) Even though the IUPAC (International Union for Pure and Applied Chemistry) has advised using the symbol *e* (for the electron) without its negative charge as superscript, we have circumvented this rule for didactic purposes since, in our experience, students are less confused when balancing charges in redox equations when they *actually* see the minus sign of the electron charge.

(b) Physical states are written here as subscripts just below the participants in chemical reactions, except for aqueous species. A few years ago, such physical states started to be written as normal letters rather than subscripts; however, we use the traditional convention here for the sake of clarity. In addition, following the usage set forth by perhaps the most referenced book worldwide in Aquatic Environmental Chemistry (Stumm and Morgan), we adhere in this text to the practice that aqueous species are to be understood as such, even when they appear without the corresponding physical state as subscript. This undoubtedly improves the readability of a large number of reactions. A similar idea applies to the gases in Chapter 4, where it would be cumbersome to write their physical states.

(c) Some chapters and subjects lend themselves more naturally than others to exercises (shown as examples).

(d) All the experiments refer to specific chapters from the theoretical section, as written below each experiment's title.

(e) Equations and figures in the worked examples are not numbered, unless they need to for a practical reason.

(f) A plethora of Environmental Chemistry experiments are performed throughout the world. As this book presents merely a few of these subjects, each chapter also has references to other experimentation found in the literature. The reader is encouraged to inform the authors of any omissions, so that future editions may be as complete as possible.

(g) Environmental Analytical Chemistry is of the utmost importance in understanding a large number of environmental issues, and there are many excellent textbooks and laboratory manuals dealing with this area of interest. Unfortunately, this rather extensive subject requires a stronger background than that assumed for readers of the present books. Furthermore, the methods and techniques involved often call for somewhat sophisticated equipment not available in all schools. It is for these reasons that we have chosen to emphasize other aspects in the present texts, and thus urge the readers to seek key references in this field elsewhere. With such a thought in mind, a comprehensive list of Environmental Chemistry experiments that give prominence to analysis—and that require instrumentation beyond that used in the experimental book—is given in the Appendix.

Jorge Ibanez first conceived the idea for this book. Zvi Szafran (New England College, USA) induced us into making this project combine a full textbook and a laboratory manual. Margarita Hernandez was the architect and Jorge Ibanez the main driving force behind the project—they weaved the threads from the different chapters into an orderly whole. In addition, Carmen Doria endowed these books with her expertise in the Life Sciences and Green Chemistry, Arturo Fregoso in the Atmospheric Sciences, and

Mohan Singh in Microscale Chemistry. In addition, all the authors participated in and reviewed other chapters as well.

Work on the books greatly benefited from comments and suggestions made by Hugo Solis (Universidad Nacional Autonoma de Mexico–Universidad Autonoma Metropolitana – Azcapotzalco, Mexico) and Mario Avila (Ecole Nationale Superiure de Chimie de Paris, France – Universidad de Guanajuato, Mexico). In addition, Dara Salcedo (Massachusetts Institute of Technology, USA – Universidad Autonoma del Estado de Morelos, Mexico), Pedro F. Zarate-Del Valle (Universite Pierre et Marie Curie, France – Universidad de Guadalajara, Mexico), Sergio Gomez-Salazar (Syracuse University, USA – Universidad de Guadalajara, Mexico), Martin Adolfo Garcia-Sanchez (ITESO – Guadalajara), and Lorena Pedraza-Segura (Universidad Iberoamericana) also helped reading some portions.

Andrea Silva-Beard gave the books the final administrative "push" for their completion. Rosa Maria Noriega provided the magic touch to the use of language through editing English grammar and style in most of the manuscript. Aida Serrano, Patricia Hernandez-Esparza, Marcela and Daniela Delgado-Velasco revised some parts of the books. Juan Perez-Hernandez (PROVITEC) helped in getting all the authors together for meetings, and Elizabeth Garcia-Pintor tested most of the experiments. Alberto Sosa-Benavides, Adriana Canales-Goerne, Gabriela Castañeda-Delgado, and Alejandro Correa-Ibargüengoitia transformed many of our rough sketches into understandable figures. The cover was developed after an idea first conceived by Carmen María Tort-Oviedo (Universidad Iberoamericana).

We are especially thankful to Ken Howell (Springer) for believing in us, and for his patience and encouragement. We also thank the many other co-authors and experimenters from other institutions that participated at different stages: Zvi Szafran (Merrimack College – Georgia Technical University, USA), Ronald M. Pike (Merrimack College – University of Utah, USA), Patricia Balderas-Hernandez (Universidad Nacional Autonoma de Mexico – Universidad Autonoma del Estado de Mexico), Bruce and Susan Mattson, Michael P. Anderson, Jiro Fujita, and Trisha Hoette (Creighton Jesuit University, USA), Alejandro Alatorre-Ordaz (Universidad de Guanajuato, Mexico), Viktor Obendrauf (Graz Pedagogical Academy, Austria), Michael W. Tausch and Michael Seesing (Universität Duisburg-Essen, Germany), Rodrigo Mayen-Mondragon (CINVESTAV-Queretaro, Mexico), Maria Teresa Ramirez-Silva (Universidad Autonoma Metropolitana – Iztapalapa, Mexico), Christer Gruvberg (University of Halmstad, Sweden), Alanah Fitch (Loyola University-Chicago, USA), Adolfo de Pablos-Miranda (Institut Quimic de Sarria, Spain), and Norberto Casillas (Universidad de Guadalajara, Mexico).

Also, many thanks are in order for extensive experimental assistance in some cases and exceptional clerical help in others, to colleagues and students at Universidad Iberoamericana: Samuel Macias-Bravo, Veronica Garces-Castellanos, Sebastian Terrazas-Moreno, Rodrigo Mena-Brito, Carlos Navarro-Monsivais, Ignacio Gallo-Perez, J. Clemente Miranda-Treviño, Jose Topete-Pastor, Luis C. Gonzalez-Rosas, Iraida Valdovinos-Rodriguez, Maria Lozano-Cusi, Ana Lozano-Cusi, Enrique Lopez-Mejia, Jose A. Echevarria-Eugui, Karla Garcia, Juan Jose Godinez-Ramirez, Fernando Almada-Calvo, Leticia Espinoza-Marvan, Rosa Margarita Ruiz-Martin, Juan Jose Arrieta, Alejandro Moreno-Argüello, and Denise Salas.

Funding was—without a doubt—of paramount importance for the development of experiments through projects and research stays at diverse stages. This was kindly provided by Universidad Iberoamericana (Mexico), the International Business Office of CONACYT (Mexico), Merrimack College (USA), the National Microscale Chemistry Center (USA), the National Science Foundation (USA), the Fulbright Program of the Department of State (USA), the Linnaeus – Palme Program of the Swedish Agency for International Development (Sweden), the Alfa Program of the European Commission (Belgium), Loyola University of Chicago (USA), Creighton Jesuit University (USA), the University of Halmstad (Sweden), and the University of Guadalajara (Mexico).

Above all, we thank our families for their gift of patience and understanding during the seemingly unending and highly demanding hours that this book required.

We are convinced that among the most important issues in Environmental Science are the appreciation and knowledge of the different phenomena involved in our environment, and the on-going need to participate in its care. We are hopeful that through these books we are contributing with a grain of sand to such an end. Our environment is undoubtedly part of a greater, transcendental reality—it is in this sense that we dedicate the present books *ad majorem Dei gloriam*.

(Note: Names of the authors appear below followed by the institutions where they did graduate work, then by their present affiliations).

Margarita Hernandez-Esparza (Stanford University). Universidad Iberoamericana, Mexico City.

Ma. del Carmen Doria-Serrano (Universidad Nacional Autonoma de Mexico). Universidad Iberoamericana, Mexico City.

Arturo Fregoso-Infante (University of Missouri-Kansas City). Universidad Iberoamericana, Mexico City.

Mono Mohan Singh (St. Petersburg Institute of Technology). Merrimack College, MA, USA.

Jorge G. Ibanez (University of Houston). Universidad Iberoamericana, Mexico City.

Mexico City and North Andover, MA, USA
Spring of 2007

Contents

Experiments

Experiment 1
Water Characterization

Reference Chapter: 6

Objective

After performing this experiment, the student shall
be able to

- Measure several parameters that indicate the characteristics and differences of various types of natural water samples: surface water, groundwater
  (mineral water) and seawater.

Introduction

The parameters to be explored in this experiment
will help us determine the main differences among
water samples. They include pH, conductivity, chloride and sulfate concentration, and hardness level
(as measured by the total amount of calcium and
magnesium ions). Besides showing the pH, these
parameters reveal the salt content of each sample,
which normally varies depending on the source. For
example, the main differences between surface and
groundwater may lie in their salt content and turbidity, which is an indirect measure of suspended
solids. The characteristics of surface and groundwater depend mainly on the nature of the catchment
area, the type of soil present, and the materials in the
confinement rock that retains the aquifer. In contrast,
the characteristics of seawater are more constant and
well known.

Experimental Procedure

The method for this experiment is to perform, on
different samples, sequential measurements of the
parameters discussed above.

The first step is to obtain samples. Preferably, you
should obtain samples from original sources, such as
river or lake water. For groundwater, take a sample
from a well if possible, but if this is unfeasible, use
a sample of bottled mineral water (preferably from
a natural source). Seawater is an ideal sample, if it is
available nearby; if it is not, prepare it synthetically
or obtain it from a commercial source.

Samples must be collected in clean polyethylene
bottles and *analyzed immediately*. This is manda-
tory for pH. For the other parameters, if immediate
analysis is not possible, refrigerate the samples at
4°C and analyze within 48 hours.

Next, measure the **conductivity/salinity** and **pH**
of the samples and compare the values with those for
the same parameters of tap and distilled or deionized
(D.I.) water. The conductivity measurements can be
done with a conductivity meter, which in some cases
is equipped for reporting the percentage of salinity
in the sample as well.

The **pH** of the samples can be measured with
a portable potentiometer or pH meter, preferably
on the sampling site, because a maximum of 2 h
is advisable for this measurement. Salinity (or an
indirect measurement, such as the **chloride ion**

concentration obtained by titration), provides interesting data for sample comparison.

Another parameter that will be measured and that also affects salinity values is the concentration of **sulfate** ions.

Total hardness, which helps to differentiate samples, as well as the amount of hardness due to **calcium** and **magnesium** ions in each sample, can be determined either by titration or by atomic absorption.

A. pH and Conductivity Measurements

Estimated time required: 1 min per sample

Safety Measures

No special precautions are needed with these measurements because a measuring probe is introduced directly into the sample and the sample is not modified.

Materials	Reagents or samples
Conductivity meter	Take river or lake water samples. Do the same with seawater, bottled commercial drinking mineral water, and —when available, use also samples of groundwater extracted from wells. (*Note:* for synthetic seawater, prepare a solution with commercial sea salt mix for seawater aquariums)
pH meter with a small-diameter electrode (for test tubes) 50-mL beakers	

Experimental Sequence

Calibrate the conductivity and pH meters. Place each sample in a clean beaker.

1. Measure the conductivity/salinity percentage of the samples by introducing the conductivity probe. Rinse the probe with D.I. water between samples, and collect the rinses in a separate beaker.
2. Measure the pH of the samples by introducing the pH probe. The pH meter must be set up and ready in advance.

Note: When not in use, the pH meter must remain in STANDBY mode and the probe bulb must stay submerged in an appropriate solution (typically a pH

7 buffer, or a KCl solution). The pH probe must be rinsed perfectly between measurements and before storage. Blot dry it before each measurement.

B. Chloride Concentration by Titration Applying the Mohr Method

In this part of the experiment the student will measure the chloride ion concentration in the water samples by promoting the formation of a white silver chloride precipitate, with the chromate ion as indicator of the endpoint (signaled by the formation of a red-brownish silver chromate).

Estimated time required: 15 min per sample

Safety Measures

Keep the titrant and the indicator from coming into contact with the skin or eyes because silver nitrate produces brown spots on the skin and the chromate indicator is toxic. All the residues generated in this experiment must be collected in a heavy metal residue bottle.

Materials	Reagents
1 microburet (i.e., a 2-mL graduated pipet in 1/100, with a syringe connected to it by means of latex or Tygon® tubing)	– Potassium chromate indicator (dissolve 0.05 g of K_2CrO_4 in 10 mL of D.I. water)
2 2-mL volumetric pipets	– 0.01 N $AgNO_3$ solution
1 propipet or a syringe (e.g., a 3-mL syringe with latex or Teflon tubing)	– Standard 0.01 N NaCl solution
2 25-mL or 10-mL Erlenmeyer flasks	– Distilled or D.I. water
6 50-mL beakers (for the sample, for the titrant and for the buret rinsing)	– Activated carbon
1 support and clamps for the microburet	– $CaCO_3$
1 Beral pipet	
1 wash bottle with distilled or D.I. water	
1 25-mm plastic filter holder	
1 10-mL syringe	
1 wash bottle with D.I. water	
0.7 μm Nitrocellulose filter membranes (25 mm diameter)	
pH meter or pH indicator paper	
1 thin spatula	
1 bottle for residues	

Experimental Sequence

1. Secure the microburet in the stand with a clamp. Pour a small amount of standardized 0.01 N silver nitrate solution (see the *Note below) into a beaker and from this, take a small amount with the microburet (using the syringe); rinse the buret. Repeat this step with another *small amount* of the solution. Collect the rinses in a separate beaker. Fill the microburet with the same silver nitrate titrant up to the 2 mL mark.
2. If the sample is colored and shows high turbidity, add a small amount of activated carbon and pass the sample through a 0.7 μm nitrocellulose filter. Use a clean, 10-mL plastic syringe to draw

$$\text{Conc. Cl}^-, \text{mg/L} = \frac{(\text{V titrant, mL}) \, (\text{Weight-equiv. of Cl}^-) \, (\text{Normality of AgNO}_3) \, (1000 \text{ mg/g})}{(\text{V sample, mL})}$$

in the sample. Connect the plastic syringe to the filter holder containing the filter membrane, and let the liquid flow through by pushing the syringe plunger softly. Collect the filtered sample in another beaker. Filter approximately 20 mL of the sample (colorless and free of suspended solids).

3. Measure the pH of the sample (using either pH indicator paper or the pH meter). If the pH is below 5, add a small amount of sodium carbonate and swirl gently before the next step. If the pH values are strongly basic (>10), neutralize first with dilute sulfuric acid and then add some sodium carbonate.
4. Measure 2 mL of the water sample (previously filtered if necessary) with a volumetric pipet, and put it in a 25-mL or 10-mL Erlenmeyer flask. Add two to three drops of the chromate indicator and swirl. Observe the color. It must be greenish-yellow.

5. Titrate drop wise, swirling the flask gently until the sample turns a pink-yellowish color. Note that a grayish precipitate forms, and at the end it takes on the color of the indicator. Record the volume of titrant used. Place the residues in the corresponding bottle, together with the rinse water used to wash the flasks.
   Repeat the process with each water sample.
6. All the residues generated in the experiment must be collected in the bottle labeled for that purpose and disposed of according to local regulations.

The concentration of chloride ions in the sample is calculated with the following equation:

C. Sulfate Concentration Applying the Turbidity Method

In this part of the experiment the student determines the presence of sulfate ions in water by observing the generation of a barium sulfate precipitate. Then the turbidity produced in the sample is measured and related to the sulfate concentration through a calibration plot. The minimum detection limit with this method is 1 mg/L of sulfate ion.

Estimated time required: 35 min

Safety Measures

Barium chloride is toxic and must not come into contact with the skin nor be inhaled. The residues from the experiment must be collected and deposited in a bottle labeled for disposal.

*Note: To standardize the silver nitrate solution, put a 2-mL sample of the 0.1 N standard NaCl solution in an Erlenmeyer flask and repeat the procedure outlined above. The volume of titrant used for titration of the standard sample will allow us to calculate the real concentration of the silver nitrate solution. Remember that at the equivalence point, the number of equivalents of the analyte equals the number of equivalents of the titrant. Use the equation below for this calculation:

$$\text{Normality of AgNO}_3 \text{ titrant} = \frac{(\text{V, mL of NaCl std. sample}) \, (\text{Normality of the NaCl std.})}{(\text{mL of the AgNO}_3 \text{ titrant used})}$$

Materials	Reagents
2 2-mL, graduated pipets (1/100)	– Na$_2$SO$_4$ standard
5 10-mL flat-bottom tubes or	solution (150 mg/L).
vials equipped with a cap and	Note: 1 mL of this
a micro magnetic stirrer	solution is equivalent to
1 propipet or a syringe adapted	100 micrograms of
with latex or Tygon tubing	sulfate ion
1 1-mL graduated pipet	– BaCl$_2$ (solid)
1 Beral pipet	– Conditioning reagent*.
1 10-mL beaker	
1 spectrophotometer set at	
420 nm	
2 spectrophotometer cuvettes	
(5-mL or smaller)	
1 5-mL syringe	
0.7 μm Nitrocellulose filters	
(25 mm diameter)	
1 25-mm plastic filter holder	
1 wash bottle with D.I. water	
6 50-mL beakers	
pH meter or pH indicator paper	
1 thin spatula	
magnetic stirring plate	
1 bottle for residues	

*Dissolve 5 mL of glycerin in 3 mL of concentrated HCl and 30 mL of distilled water; add to this mixture, 10 mL of isopropyl alcohol and 7.5 grams of sodium chloride. Mix everything until perfectly dissolved. **This reagent must be prepared in advance.**

Experimental Sequence

A. Prepare **at least five dilutions** of the sulfate standard solution. Apply the technique described below to analyze each dilution, and generate the data to build the calibration curve. Because the total volume of sample used with this method is 4 mL, prepare the following dilutions of the standard:

- 4 mL of standard = 100%
- 2 mL of standard + 2 mL of D.I. water
- 1 mL of standard + 3 mL of D.I. water
- 1.5 mL of standard + 2.5 mL of D.I. water
- 0.5 mL of standard + 3.5 mL of D.I. water
- 4 mL of D.I. water as a blank.

All of these dilutions must undergo the entire procedure that applies to the water samples.

All the water samples must be pre-filtered to free them of any solids.

B. *Use the following technique with each dilution of the standard as well as with each water sample.*

1. With a pipet, pour exactly 4 mL of sample into a vial with a magnetic microstirring rod and cap.
2. With the graduated pipet, add 0.6 mL of the conditioning reagent and stir magnetically. (*Note*: The conditioning reagent acidifies the medium in order to favor the precipitation reaction and to eliminate the possibility of precipitation of the barium carbonate that may form in highly alkaline waters. The glycerin favors the dispersion of the colloidal precipitate formed in the liquid medium, allowing a better turbidimetry measurement).

   Add a small amount of barium chloride with the spatula. It is preferable to add an excess of this reagent, in order to favor the common ion effect and to accomplish the complete precipitation of the sulfate ions. Stir constantly.
3. Keep stirring a minute longer; then allow the mixture to stand for 2 minutes. Watch for any turbidity that may form. A turbidity lasting several minutes (while stirring) signals the presence of sulfate ions.
4. Turn on the spectrophotometer and set it at 420 nm. Allow it to warm up. Calibrate to 100% transmittance with a spectrophotometer cuvette containing D.I. water.

   Any barium-treated sample that shows turbidity must be well mixed and poured into a spectrophotometer cuvette and *inserted into the spectrophotometer*. Immediately afterwards, read the transmittance or the absorbance *continuously at 30-second intervals for 4 minutes*.

   The maximum absorbance (turbidity) is generally obtained within 3–4 minutes, and sometimes this value lasts for one minute or longer. This maximum value is used for calculating the sulfate concentration.

   The experimental conditions must be the same for all samples in order to obtain reliable values.
5. To determine the sulfate concentration of the water sample, you must first plot the calibration curve with the *data of absorbance vs the corresponding concentrations* (i.e., from the dilutions of the standard ion). Once you know the absorbance value of the sample, you can calculate the concentration of sulfate ions in the water sample.
6. All the residues generated in the experiment must be collected in the bottle labeled for that purpose.

D. Total Hardness, Calcium, and Magnesium Ion Concentrations

The student will determine total hardness as well as the calcium and magnesium ion content in a water sample applying the EDTA titration method.

Estimated time required: 10 min per sample

Safety Measures

In this case, the student must proceed with caution in handling the sodium hydroxide for the pH adjustment. In the event of a spill or skin contact, wash with abundant water. The residues generated can be neutralized and flushed down the drain.

Materials	Reagents
1 microburet (i.e., a 2-mL graduated pipet in 1/100, with a syringe connected to it by means of latex or Tygon® tubing)	2 M NaOH solution (for pH adjustment)
2 small spatulas	Solid murexide indicator
1 2-mL volumetric pipet	Solid eriochrome black indicator
1 2-mL graduated pipet (in 1/100)	0.01 M or 0.001 M Na₂EDTA (sodium ethylenediamine tetra acetate) solution
1 propipet or a 3- or 5-mL syringe (adapted with a small latex or Tygon® tube)	
6 25- or 10-mL Erlenmeyer flasks	pH 10 buffer (NH₃ / NH₄⁺)
4 10-mL beakers	
1 pH meter equipped with a small-diameter combination pH electrode (for test tube insertion), or pH indicator paper	
1 stand and a clamp for the microburet	
2 Beral pipets	
1 wash bottle with distilled or D.I. water	

Experimental Sequence

1. Measure the pH of the sample with a pH meter or pH indicator paper. Fill the microburet with the concentrated EDTA solution, and adjust to a known volume.

2. In order to measure the total hardness value (i.e., the $Ca^{2+} + Mg^{2+}$ concentration), place a 2-mL, solids-free water sample (measured with a volumetric pipet) in an Erlenmeyer flask. Add 2–3 mL of the pH 10 buffer, swirl and add one or two crystals (or a small amount of powder) of the Eriochrome black solid indicator. Swirl until total dissolution. The mixture should now appear with a red wine color. Titrate this with the EDTA solution to a dark-blue endpoint. If the amount of titrant needed to reach the endpoint is too small to be measured, repeat the titration with another sample using a more dilute titrant (e.g., 0.001 M EDTA). Note: If the blue color appears from the start, this means there is no measurable hardness in the sample.

3. With a volumetric pipet, put 2 mL of the sample (free of solids) in a 25-mL Erlenmeyer flask. Add 1 mL of 2 M NaOH to ensure that the pH is frankly basic (pH $\sim$ 11); add one or two crystals of solid murexide indicator (or a small amount of its powder), and swirl softly until they dissolve. Titrate with 0.01 M EDTA to a violet endpoint. If the amount of titrant needed to reach the endpoint is too small to measure, repeat the titration with another sample using a more dilute titrant (e.g., 0.001 M EDTA). This value will let us know the calcium ion concentration in the sample. Repeat this method with each water sample.

4. One mole of EDTA is consumed for each mole of Ca^{2+} or Mg^{2+}. Because the MW of calcium carbonate is virtually equal to 100, then the concentration of Ca or Mg (expressed as mg/L of calcium carbonate) can be calculated with the following equation:

Total or Ca^{2+} hardness, mg/L as $CaCO_3$

$$= \frac{(V_{titrant} \times \text{EDTA Molarity} \times 100 \text{ g/mol} \times 1000 \text{ mg/g})}{V_{sample}}$$

Name_____Section_____Date_____

Instructor_____Partner_____

PRELABORATORY REPORT SHEET—EXPERIMENT 1

Experiment Title _____

Objectives

Procedure flow sheet

Waste containment procedure

PRELABORATORY QUESTIONS AND PROBLEMS

1. Explain why the indicated parameters need to be measured so as to differentiate the three types of natural waters.

2. Explain what precautions must be taken when sampling a natural water source.

3. Explain why the pH must be measured on-site, whereas measurement of the other parameters requires that the samples be refrigerated.

4. Why are polyethylene bottles rather than glass bottles preferred for storing water samples?

5. Write down the reactions that correspond to the measuring technique for

    a. Chloride ions
    b. Sulfate ions
    c. Calcium ions, and total hardness

6. What causes the conductivity of water? What are the common units for this parameter?

7. What is salinity? What ions contribute to its value? What salinity values are common in seawater? In groundwater?

8. Why is hardness (as well as calcium and magnesium hardness) reported as ppm of calcium carbonate?

9. Why is it necessary to make the pH basic in order to measure the calcium ion concentration?

10. What would happen in the case of the total hardness EDTA titration if the pH were not adjusted to 10 with the buffer?

11. In what kind of water sample would you expect to find higher hardness values and why?

Additional Related Projects

• Use the above procedures with other water samples (e.g., tap water, D.I. or distilled water, rain water) and compare the values for the measured parameters.

Name_____Section_____Date_____

Instructor_____Partner_____

LABORATORY REPORT SHEET—EXPERIMENT 1

Observations

Origin of sample

#1_____

#2_____

#3_____

Observable characteristics of the sample (color, odor, suspended solids present, etc.)

#1_____

#2_____

#3_____

Precautions or measures taken during the sampling. Sampling procedure for each sample

#1_____

_____

#2_____

_____

#3_____

_____

A. pH and conductivity

Sample number   pH of the sample

#1_____ _____

#2_____ _____

#3_____ _____

Sample number   Conductivity/percentage of salinity

#1_____ _____

#2_____ _____

#3_____ _____

B. Chloride concentration

B.1. From the experimental data of each sample tested, report the true normality of the titrant.

Normality of the silver nitrate titrant

Volume of the sodium chloride standard solution: _____mL

Concentration of the sodium chloride standard solution: _____N

Volume of the silver nitrate titrant: _____mL

True concentration of the silver nitrate titrant: _____N

Equation:

B.2. Calculate and report the chloride concentration in each sample as mg/L.

Chloride concentration:

| Sample number | Sample volume, mL | Vol. of silver nitrate titrant, mL |

#1_____   _____   _____

#2_____   _____   _____

#3_____   _____   _____

Sample number   Chloride ion concentration, mg/L

#1_____   _____

#2_____   _____

#3_____   _____

Equation:

C. Sulfate ion concentration

Data for the standard calibration curve:

Sulfate standard concentration: _____

Standard dilutions considered:

No.	Standard volume / 4 mL	Sulfate concentration	Max. absorbance
1	4/4		
2			
3			
4			
5			
6			

Sulfate calibration curve: Plot the sulfate concentration vs. absorbance and include as an Appendix.

Equation of the calibration curve:

SO_4^{2-}, mg/L =

Correlation coefficient: $r^2 = $ _____

Experimental absorbance values

Sample #1:_____

Time, s	Absorbance
0	
30	
60	
90	
120	
150	
180	
Maximum value:	

Sample #2:_____

Time, s	Absorbance
0	
30	
60	
90	
120	
150	
180	
Maximum value:	

Sample #3:_____

Time, s	Absorbance
0	
30	
60	
90	
120	
150	
180	
Maximum value:	

Sulfate ion concentration in each sample:

Sample number   Sulfate ion concentration, mg/L

#1_____   _____

#2_____   _____

#3_____   _____

D. Hardness concentration

D.1. From the experimental data of each sample tested, report the titration values for the calcium ion and for the total hardness determination.

Calcium ion concentration

Sample #	Sample volume, mL	EDTA concentration, M	EDTA titrant volume, mL	Ca^{2+}, mg/L

Total hardness concentration

Sample #	Sample volume, mL	EDTA concentration, M	EDTA titrant volume, mL	Hardness as $CaCO_3$, mg/L

D.2. Report the mg/L of calcium carbonate for total hardness, and the calcium ion concentration in each sample. Calculate (by difference) the magnesium ion concentration, expressed as mg/L of $CaCO_3$.

Sample #	Hardness as $CaCO_3$, mg/L	Calcium as Ca^{2+}, mg/L	Calcium as $CaCO_3$, mg/L	Magnesium as $CaCO_3$, mg/L	Magnesium as Mg^{2+}, mg/L

D.3. Organize the water samples tested in decreasing order of total hardness. Explain the results according to the nature or origin of the sample.

E. Parameter comparison

E.1. Classify the water samples tested in decreasing order of the following parameters:

pH	Conductivity	Salinity	Chloride conc.	Sulfate conc.

E.2. Look at the parameters measured in these samples and observe the differences in their values as well as their order. Indicate the main effects resulting from the composition of the samples.

E.3. If possible, calculate the dispersion of the average obtained for each determination in this experiment and comment on the values achieved.

Student Comments and Suggestions

Literature References

Sawyer, C. N.; McCarty, P. L.; Parkin, G. F. *Chemistry for Environmental Engineering*, 4th ed.; McGraw-Hill: New York, 1994.

Szafran, Z.; Pike, R. M.; Foster, J. C. *Microscale General Chemistry Laboratory with Select Macroscale Experiments*; Wiley: New York, 1993.

Experiment 2
Dissolved Oxygen in Water

Reference Chapters: 6, 7, 8

Objectives

After performing this experiment, the student shall be able to:

- Determine the level of dissolved oxygen in a sample of water using Winkler's method.
- Analyze the effects of various factors on the level of dissolved oxygen in a water sample (e.g., salt content, temperature, degree of mixing, and the presence of reducing compounds).

Introduction

The level of dissolved oxygen in water is one of the most important parameters in determining its quality, because it indirectly indicates whether there is some kind of pollution. Common processes that pollute surface waters include the discharge of organic matter derived from municipal sewage or industrial wastes, and runoff from agricultural lots and livestock feedlots. In addition, the release of warm or hot discharges from industrial cooling towers induces what is known as thermal pollution. Such discharges directly affect the level of dissolved oxygen in water bodies, which is crucial for the survival of aerobic organisms and aquatic fauna such as fish; in fact, excessive pollution has caused massive fish deaths. In the long run, the discharges of organics or of nutrients favor the accelerated eutrophication or productivity process with algal blooms. As a consequence, there will be a lowering of the *dissolved oxygen content* (or *DO level*) and the "death" of the aquatic system. (See Section 8.4.2).

The measurement of the DO is also important to determine whether a water system is predominantly aerobic or anaerobic, predict the survival of aquatic organisms, and predict whether aerobic biological processes can take place for transforming the biodegradable organic contaminants discharged in water. When there is an organic discharge, the DO decreases rapidly due to the action of the aerobic microorganisms that consume oxygen during the metabolic degradation of organic matter (see Chapter 7). Consequently, the presence of dissolved oxygen is critical for the self-cleansing of the water system, and in combination with the presence of CO_2, it is also critical for the determination of the corrosive character of water on materials such as iron and other metals.

The DO depends on water temperature, dissolved salts, atmospheric pressure (and therefore of altitude), the presence of reducing compounds, suspended matter, and living species. The aquatic flora and fauna can contribute to the consumption or to the production of oxygen in water. For the aquatic flora, the photosynthesis process is responsible for the generation of oxygen in the presence of light; this explains the DO fluctuations during the day/night cycles and in the different seasons.

The presence of microorganisms and biodegradation processes affect the DO level as well. This can also vary vertically, decreasing in the water column in lakes, large deep rivers, or the ocean due to stratification. By contrast, in rivers and streams it can vary horizontally along the course of the river flow, increasing where there are waterfalls or rapids, and decreasing in the slow-moving portions of the river and in those with organic discharges or microbial activity. Since the dissolution of air in water is an interface mass-transfer phenomenon, the degree of contact and of mixing with water is also important.

The content of oxygen in air is only 21% (v/v) and it dissolves in water according to the following equation:

$$O_{2(g)} \rightleftarrows O_{2(aq)} \qquad (1)$$

This can be represented by Henry's law (see Section 6.2.1):

$$K_{H,O_2} = M_{O_2,w}/p_{O_2} \qquad (2)$$

where

K_{H,O_2} = Henry's constant for oxygen

$\qquad = 1.29 \times 10^{-3}$ mol/L-atm (at 298 K)

$M_{O_2,w} = [O_{2(aq)}]$ = concentration of the dissolved

$\qquad$ oxygen in water, in mol/L

$p_{O_2(g)}$ = partial pressure of oxygen in air at

$\qquad$ ambient temperature, in atm

Because the partial pressure of oxygen is approximated as $p_{O_2(g)} = 0.21\ P_{atm}$ (in atm), one can determine its concentration in water (in ppm) with this relationship by multiplying it times the MW of oxygen by 1000 (i.e., 32,000):

$$M_{O_2,w} \text{ (in mg/L)} = (K_{H,O_2})(p_{O_2(g)})$$
$$= (K_{H,O_2}) \times 0.21 \times 32,000 \quad (3)$$

Henry's constant for dilute solutions may approach the real equilibrium constant for the dissolution reaction; however, as the ionic strength of the solution increases, this is no longer correct, and activities or effective concentrations must substitute concentrations in the equilibrium ratio. Since oxygen is a non-electrolyte, it undergoes a salting-out effect where the water molecules bind to salts, reducing the solubility of the gas.

One of the two common methods for measuring the DO in water involves an amperometric method based on a selective electrode for oxygen, which is the most direct type of measurement. The other, more traditional method involves a volumetric titration—it is known as Winkler's method and consists in the oxidation of Mn^{2+} ions by the dis-

solved oxygen under basic conditions. This step of the reaction is called the *oxygen-fixation*. If there is no oxygen present, this step generates a white precipitate of $Mn(OH)_2$; if there is oxygen present, a brown precipitate of $MnO(OH)_2$ and MnO_2 is formed. This precipitate is dissolved with an acid, and when iodide ions are present, they are oxidized to molecular iodine, I_2, which gives a brownish color to the solution. The amount of liberated iodine is directly related to the amount of dissolved oxygen. This iodine is titrated with a standardized sodium thiosulfate solution, using a starch indicator for the endpoint. The presence of nitrites can interfere and affect the results noticeably; however, if sodium azide is added, it transforms the nitrites into N_2 and N_2O. The addition of azide is known as the *modified Winkler* method.

In this experiment, students will take several samples of water—some of them saturated with air—and measure the DO using Winkler's method, in a semi micro scale adapted from an original method proposed by Ondrus (see Ondrus, 1993).

Because the mixing of reactants must be done without introducing additional air to the sample, a 10-mL syringe is used (instead of a standard DBO bottle). In this way, the sampling, the addition of reactants, and the reaction all take place in the syringe. Another option is to use a small vial of known volume, equipped with a septum seal, and add the reactants to it with a syringe. At the end, the sample is deposited in a small Erlenmeyer flask and titrated to the endpoint.

The temperature, the mixing and salt content of a water sample are to be modified as well, so as to allow visualization of the factors that affect the dissolution and the concentration of oxygen in water. Finally, to observe the depletion of oxygen by the addition of a reducing compound, one can measure the DO before and after reaction with a reducing agent. The DO is measured with the same method described above or with the selective electrode method.

Experimental Procedure

Estimated time to complete the experiment: 15 min per sample

Materials	Reagents or samples
1 10-mL plastic syringe with its extreme adjusted with a plastic pipet tip or a piece of Tygon® tubing and a valve; or 1 10-mL bottle adapted with a septum seal	0.22 M $MnSO_4$ Alkaline 0.3 M KI and sodium azide solution. (Dissolve 5 g of KI, 2.4 g of NaOH and 0.3 g of NaN_3 made up to 100 mL)
1 3- or 1-mL syringe with the needle cut	0.018 M H_2SO_4 (dissolve 1 mL of pure acid to a total
1 100-mL volumetric flask	of 100 mL with D.I. water)
2 50-mL volumetric flasks	0.00025 M freshly-prepared
2 50-mL Erlenmeyer flasks	standard $Na_2S_2O_3$ solution
1 Beral pipet	sodium metabisulfite,
1 25-mm filter holder	$Na_2S_2O_5$ (solid)
1 10-mL syringe	starch indicator solution
6 25-mL beakers	Na_2SO_3 (solid)
0.7 μ 25-mm Nitrocellulose	NaCl (solid)
filters	$CoCl_2$ (solid)
1 microburet (e.g., a 2-mL 1/100 graduated pipet with a 3- or 5-mL syringe adapted with latex or Tygon® tubing; or an insulin syringe adapted with a thin Tygon® tubing, see Figure I)	tap water D.I. water river or lake water
1 wash bottle with D.I. water	
3 air pumps (e.g., aquarium air pump) equipped with Tygon tubing and a diffuser	
5 100-mL beakers	
1 support stand and clamps for the microburet	
1 ice bath	
3 thermometers or thermocouples	
2 hot plates with sand bath	
1 stir plate	
1 magnetic stirrer	
1 small spatula	
1 25-mL graduated cylinder	
1 bottle for residues	
1 10-mL graduated pipet	
1 weighing balance	

Safety Measures

Sulfuric acid is corrosive and contact with the skin or eyes must be avoided. All the residues generated in this experiment must be disposed of in a special container identified for that purpose.

Before proceeding with the experiment, obtain the value of the current barometric pressure.

Note: An incorrect sampling process in natural waters can alter the results considerably due to the in-

troduction of extra-dissolved oxygen. In order to sample correctly, take the sampling bottle or vial with a gloved hand and introduce it slowly and horizontally down the water surface to the height of the sampling site. Rinse the bottle with water under the surface, then fill it completely and cap it inside the water or immediately after it reaches the surface; make sure that there are no air bubbles trapped inside. If samples from deeper levels are needed, a special sampler must be used.

Part A. Measurement of the Dissolved Oxygen Level (D.O.) of an Air-Saturated Sample of D.I. Water and of a Sample of Tap Water

1. Place approximately 50 mL of D.I. water in a 100-mL beaker and saturate it with air by aeration with an air pump through a diffuser, for at least 20 minutes. Write down the ambient temperature.
2. In the meantime, apply Winkler's method to a tap water sample. This technique will be used throughout the experiment, with all samples.
3. After aeration, determine the D.O. level in the saturated D.I. sample using the same technique.

Steps in the Microscale Winkler's Technique

1. Rinse the 10-mL syringe prepared as shown in Figure 1 by pulling the plunger and slowly drawing approximately 2 mL of the water sample. Then, turn the syringe upwards and tap the body of the syringe softly so as to liberate the small bubbles on the surface of the plastic material.

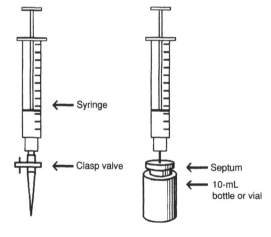

FIGURE 1. Adapted syringe or vial for reaction.

Let them buoy to the tip and press the plunger to release them from the syringe. Now turn the syringe downwards and slowly push the plunger to expel the entire sample. This procedure will leave a small amount of water filling the tip of the syringe. If the plunger is then softly pushed further, a small drop of water will form at the tip of the syringe.

2. Now take approximately 6 mL of the target water by introducing the tip of the syringe deep inside the water sample in the beaker, and slowly pulling out the plunger. No visible air bubbles, except those of the sample, must be contained in the syringe.

3. Next, incline the syringe tip upwards and slowly push the plunger in, so as to expel the necessary volume and leave exactly 5 mL of sample inside. If a vial is used instead of the syringe, fill it to capacity (record the exact volume added), and seal it with a septum.

Care must be taken in order to prevent any air bubbles from getting into the syringe or vial as the following reactants are introduced:

a. Draw 0.5 mL of the manganese sulfate solution into the syringe or vial (hereafter called the *reaction chamber*). Take care not to introduce more than what is specified, because once the solution is inside, it cannot be taken out. If you use a syringe, close its tip using a septum, or by either bending or clasping tightly a piece of tubing attached to the tip. If you use a vial, introduce the reactant with the syringe. In either case, after that, shake the system up and down several times to mix the contents well.

b. Draw 0.5 mL of the alkaline iodide-azide solution into the syringe or vial in the same manner as in the previous step. The plunger tip must now be at the 6 mL mark. Once more close the tip of the syringe and mix thoroughly as mentioned above (i.e., allow it to react mixing the contents by shaking the reaction chamber up and down at least 10 times or for 1 or 2 minutes). Observe the kind of precipitate formed. If it is cloudy-brownish, this signals the presence of manganese hydroxide (III), which also indicates that there is dissolved oxygen present. If it is a cloudy-white precipitate, then this means that no dissolved oxygen is present.

c. Carefully draw 2–3 mL of 0.018 M sulfuric acid into the reaction chamber in the same manner described above. Shake the entire reaction mixture carefully but thoroughly until the precipitate dissolves in the acid medium. If a small residue of precipitate remains, add a very small amount of extra acid so as to dissolve it.

(Note: After adding sulfuric acid, the presence of air in contact with the sample will not affect the results).

The color of the final solution (from pale yellow to brown) will depend on the amount of dissolved oxygen initially present in the sample. The darker the color, the higher the amount of initial oxygen dissolved. A brown color reflects the presence of I_2^0 (in the form of I_3^-).

Once the reaction is complete (i.e., the precipitate is dissolved), dispense the content of the syringe into a clean, 50-mL Erlenmeyer flask. Use 3–5 mL of D.I. water to rinse the body of the syringe and add it to the sample in the flask. Fill a microburet with the 0.00025 M $Na_2S_2O_3$ solution and titrate the iodine in the sample (liberated as I_3^-) to a yellowish color. Then add two to three drops of starch indicator. A green to blue color must develop. Softly, swirl the Erlenmeyer flask and continue titrating slowly to a colorless endpoint. Record the total volume of titrant used.

Part B. Dissolved Oxygen Measurement of an Aerated D.I. Sample Pre-treated with a Reducing Agent

Take a sample of 20 mL of the aerated D.I. water, place in a beaker, and with a thin spatula add a very small amount of a reducing agent, that is, sodium sulfite or sodium metabisulfite, and a small crystal of cobalt (II) chloride as catalyst. Let the mixture react, then take a filtered sample of the treated water and apply the same technique as before to measure the dissolved oxygen. Observe the result after the reducing treatment, and compare it with the D.O. of the original sample (i.e., aerated D.I. water).

Part C. Effect of Temperature in the Dissolved Oxygen Level in Water

Prepare three different samples of D.I. water of 50 mL each in a beaker, and place them in different temperature baths: for example, (a) one in an ice bath (~4°C), (b) a second one at 40°C and, (c) another at 80°C. These last two may be prepared in a sand bath, each at different temperature. With the air pump let air bubble into the separate samples in their corresponding baths, for at least 10 minutes each. After this, determine the temperature, take a sample of each, and measure the dissolved oxygen level using the modified Winkler technique.

Part D. Effect of Ionic Strength in the Dissolved Oxygen Level in Water

Because the ionic strength or salt content will affect the dissolved oxygen level, prepare 100 mL of a blank solution containing 50 g/L of NaCl. Prepare three beakers: (1) in one beaker, put 50 mL of the blank solution; (2) in a 50-mL volumetric flask, dilute 30 mL of the blank solution to 50 mL; (3) prepare another 50 mL in a volumetric flask with 15 mL of the blank solution diluted with D.I. water. Place each solution in a beaker and saturate with air for at least 15 minutes, and then take a sample and measure the D.O. with the same modified Winkler method.

Part E. Dissolved Oxygen Measurement of Two Samples of Surface Water (River or Lake): One Fixing the Oxygen on Site and the Other, After Transporting it to the Laboratory

1. Carry out the measurement of dissolved oxygen with the modified Winkler technique of a river or lake surface water sample. The most reliable results are obtained if the sample is processed and the dissolved oxygen is fixed through the first two reactions of the Winkler method at the site of sampling, after which the processed sample is carefully transported to the laboratory for the titration process (three samples for more reliability). The sampling may be done with a vial or directly with the syringe. That is, carry out the first three steps of the Winkler technique, and leave the titration step for the laboratory.

2. Take another surface water sample at the same site and this time do not fix the oxygen (i.e., do not add any reactants); only take care to cap the sample well until it is measured in the laboratory. Then, take the capped sample to the laboratory and carry out the Winkler technique as specified. After doing this, you will see the difference in the measurements on site and off site.

Name_____Section_____Date_____

Instructor_____Partner_____

PRELABORATORY REPORT SHEET—EXPERIMENT 2

Objectives

Procedure flow sheet

Waste containment procedure

PRELABORATORY QUESTIONS

1. Explain why the dissolved oxygen content of an environmental sample of water is important.

2. Look up the reported limit values for dissolved oxygen in water so that a system may support aquatic life. What would be the limit value for a system to still be considered aerobic? Anaerobic?

3. Explain what contributes or favors an increased value of dissolved oxygen in water.

4. Which are the main types of pollutants that affect the level of dissolved oxygen in water?

5. In what cases would it be undesirable for dissolved oxygen in water to be present?

6. Establish all the chemical reactions that take place in the modified Winkler Method.

7. Explain what is understood as the ionic force of a solution, what determines it, and how it can be calculated.

8. Explain the concept of salting out and how it affects the solubility of non-electrolytes.

9. What other concept and units besides mg/L is used to express the presence of dissolved oxygen in water? How is this calculated?

10. Using the theoretical approach, calculate the saturation concentration of dissolved oxygen under standard conditions.

11. Using the theoretical approach, explain how you would calculate the concentration of dissolved oxygen at other latitudes and at other temperatures.

Additional Related Projects

- Consider allowing the reduced sample to re-aerate and measure the D.O. level with time in order to determine when the reducing agent is consumed and how fast the saturation level is attained again.

- Consider using in the experiment samples from other natural systems, for example, ocean water.

- An additional experiment can be carried out in which an air-saturated water sample properly seeded with activated sludge is dosed, and the D.O. is measured with time.

- Another experiment would be to add a small amount of a biodegradable organic substrate to an air-saturated water sample seeded with activated sludge and to measure the level of dissolved oxygen vs. time (preferably with the amperometric method). Winkler's method has the disadvantage of being a destructive test, and therefore the amount of mixture or sample would have to be large enough to allow the measurements, yet not interfere too much with the process.

Name_____ Section_____ Date_____

Instructor_____ Partner_____

LABORATORY REPORT SHEET—EXPERIMENT 2

Part A. *Measurement of the dissolved oxygen level (D.O.) of an air-saturated sample of D.I. water and a direct sample of tap water*

Experimental data

Sample A#1: Air-saturated D.I. water

   Temperature: _____

   *Barometric pressure:*_____

   *Sample volume used:*_____

   *Color of the observed precipitate after the first two reactant additions:*_____

Sample A#2: Tap water

   *Temperature:*_____

   *Barometric pressure:*_____

   *Sample volume used:*_____

   *Color of the observed precipitate after the first two reactant additions:*_____

   Titrant:_____Titrant concentration:_____

Sample	mL of titrant	D.O. concentration, mg/L
Air-saturated D.I. water		
Tap water		

Formula used:

Part B. *Dissolved oxygen measurement of an aerated D.I. sample treated with a reducing agent*

Experimental data

Sample B#1: Aerated D.I. sample treated with a reducing agent

Temperature:_____

Barometric pressure:_____

Sample volume used:_____

Reducing agent:_____

Color of the observed precipitate after the first two reactant additions:_____

Titrant:_____ Titrant concentration:_____

Sample	mL of titrant	D.O. concentration, mg/L
B#1		

How does the value compare with the two previous samples? Was the reducing agent efficient?

Part C. *Effect of temperature in the dissolved oxygen level in water*

Barometric pressure:_____

Sample volume used:_____

Titrant:_____ Titrant concentration:_____

Sample	Sample temperature, °C	mL of titrant	D.O. concentration, mg/L
A#1			
C.a			
C.b			
C.c			

How do the values vary with temperature?

Part D. *Effect of ionic strength (or salt content) in the dissolved oxygen level in water*

Experimental data

Barometric pressure:_____

Sample volume used:_____

Titrant:_____ Titrant concentration:_____

Sample	Conc. NaCl, mg/L	Theoretical ionic strength, I (mol/L)	mL of titrant	D.O. concentration, mg/L
A#1	0	0		
D.a				
D.b				
D.c				

Formula used and calculations for the ionic strength:

Draw a graph of D.O. concentration as a function of ionic strength.

Part E. *Dissolved oxygen measurement of two samples of river or lake surface water: 1) Fixing the oxygen on site, and 2) after transporting it to the laboratory.*

Experimental data

Sample E#1: Surface water measured on site

Source:_____

Description of site and place (level) of sampling:_____

_____

_____

_____

Sampling conditions or precautions:_____

Temperature:_____

Barometric pressure:_____

Sample volume used:_____

Color of the observed precipitate after the first two reactant additions:_____

Sample E#2: Surface water measured at the laboratory

*Source:*_____

*Description of site and place (level) of sampling:*_____

_____

_____

_____

*Sampling conditions or precautions:*_____

_____

*Temperature:*_____

*Barometric pressure:*_____

*Sample volume used:*_____

*Color of the observed precipitate after the first two reactant additions:*_____

*Titrant:*_____ *Titrant concentration:*_____

Sample	mL of titrant	D.O. concentration, mg/L	% saturation with respect to saturated D.I. water
E#1			
E#2			

POSTLABORATORY PROBLEMS AND QUESTIONS

1. Does the value of the measured parameter differ from one sample to another? What may be the causes of such differences?

2. Explain how you calculated the saturation percentage of the samples.

3. Is each D.O. level acceptable for aquatic life? Why?

4. Why is it important to measure this parameter on site?

Student Comments and Suggestions

Literature References

Ondrus, M. G. *Laboratory Experiments in Environmental Chemistry*; Wuerz Publishing: Winnipeg, Canada, 1993.

Sawyer, C. N.; McCarty, P. L.; Parkin, G. F. *Chemistry for Environmental Engineering*, 5th ed.; McGraw Hill: New York, 2003.

Stumm, W.; Morgan, J. J., *Aquatic Chemistry: Chemical Equilibria and Rates in Natural Waters*; Wiley Interscience: New York, 1996.

Experiment 3
Alkalinity and Buffering Capacity of Water

Reference Chapters: 6, 8

Objectives

After performing this experiment, the student shall be able to:

- Determine the alkalinity and buffering capacity of several types of water samples: surface water, groundwater (mineral water) and sea water.
- Prepare different solutions or mixtures of acids and their conjugate bases (i.e., buffers), and measure their buffering capacity by titration with acids and bases.
- Calculate the concentrations of an acid and of its conjugate base to create a buffer for a desired buffering capacity at a specific pH.
- Prepare a buffer system.

Introduction

This experiment will allow the determination of the *alkalinity* and *buffering capacity* of water samples from different natural sources. The buffering capacity is the ability to neutralize the pH and the resistance to change in it due to the small acidic or basic inputs or discharges. When a system is poorly buffered, the addition of even small amounts of an acid or a base will noticeably alter its pH, but when a system is well buffered, the same addition barely modifies its pH (i.e., it becomes relatively insensitive to the addition of small amounts of acids or bases). The buffering capacity of a system is defined as the moles/L of strong acid (or strong base) needed for a change in one pH unit of a solution. A typical buffer is formed by a combination of a weak acid

(or base) with its corresponding salt. For example:

$$HA_{(aq)} \rightleftarrows H^+_{(aq)} + A^-_{(aq)} \qquad (1a)$$

The equilibrium (acidity) constant is:

$$K_a = \frac{[H^+][A^-]}{[HA]} \qquad (1b)$$

from which we can derive the equation of pK_a ($= -\log K_a$) with respect to the pH:

$$pH = pK_a + \log([A^-]/[HA]) \qquad (2a)$$

This is known as *Henderson–Hasselbalch equation* and it is built under the assumption that $[H^+]$ or $[OH^-] \ll [HA]$ and $[A^-]$, where HA = weak acid, and A^- = the corresponding anion generated from the salt. In a well-buffered system, the greatest resistance to changes in pH will occur when the ratio of concentrations of the acid and its salt are approximately equal and therefore the pK_a will be equal to its pH. From the above equation it is clear that this occurs at $[A^-]/[HA] = 1$.

By knowing the pK_a of the buffering acid, one can estimate the pH at which its greatest buffering capacity will be centered. The pH of a buffer solution is affected by two factors: the concentration ratio, [A-]/[HA] (i.e., the inverse ratio of the acid to the conjugate base), and the strength of the parent acid or base. The stronger the parent acid or base in the buffer solution, the more extreme will the buffer's pH value be.

The buffering capacity depends on the concentration of the buffer, and on the type and concentration of the acid or base to be added to the buffered solution. In selecting the right working buffer for a specified pH, it is common to consider that its pK_a

must be at least one pH unit above or below the working pH.

The buffering capacity in natural waters is mainly due to the carbonate system and its equilibria. Therefore, it is important to know the alkalinity of the system, because this will provide the capacity for neutralizing an acid. The expression for *alkalinity* (i.e., dissolved species only) or *acid neutralizing capacity (ANC)* (i.e., the whole sample) is generally based on the carbonate system:

$$\text{Alkalinity} = \text{ANC} = [HCO_3^-]$$
$$+ 2[CO_3^{2-}] + [OH^-] - [H^+] \quad (2b)$$

and this property is expressed as mg/L (or in eq/L, in the case of ANC) of the equivalent calcium carbonate.

The ANC of natural water systems depends on the composition of the watershed. If there are minerals with poor solubility in the surrounding soil, the ANC will be low, whereas if calcareous minerals are present, there will be a high ANC. Some dissolved organic substances derived from decaying plant materials may also contribute to the ANC capacity of the water.

In this experiment the student will use samples as those considered in Experiment 1 and determine their alkalinity and buffering capacity. The student will also prepare a series of solutions where the concentration of a known conjugate salt will vary, measure how the buffering capacity changes with the proportions of the salt, and calculate the buffering range or limits. In this case, the proportion of salt that gives the highest buffering capacity or range (and its corresponding pH) will be determined. Finally, the student will calculate the composition of a buffer solution required to give a specified pH; prepare it based on theoretical concepts and equations, and measure its pH and buffering capacity.

Experimental Procedure

Estimated time required to complete the experiment: This depends on the number of samples analyzed (approx. 5–20 min per sample per analysis).

This experiment may be carried out in one or two sessions.

The first step is to obtain water samples. The student should obtain the samples from the original sources similar to what was done for Experiment

1. Appropriate samples include river or lake water, groundwater (if possible from a well, and if not, a sample of bottled mineral water—preferably from a natural source), seawater (if not available, it can be prepared synthetically or obtained from a commercial source). The samples must be collected in clean polyethylene bottles and analyzed immediately after sampling; otherwise, the values may vary substantially.

Materials	Reagents
1 pH meter with a thin test tube pH combination electrode	mineral or groundwater
	river or lake water
polyethylene bottles	seawater
4 50-mL beakers	D.I. water
1 1-mL volumetric pipet	0.25 M, 0.1 M NaOH
2 2-mL microburet	phenolphthalein indicator
1 three-finger clamp	methyl orange indicator
2 Beral pipets	bromocresol green indicator
5 25-mL beakers	0.25 M, 0.1 M, 0.01 M HCl
2 10-mL volumetric pipets	0.01 M $Na_2S_2O_3$
1 2-mL volumetric pipet	0.1 M Na_2CO_3
3 10-mL Erlenmeyer flask	0.1 M $NaHCO_3$
2 propipet bulbs or adapted syringes	0.1 M KH_2PO_4
	0.1 M K_2HPO_4
1 universal stand	0.1 M H_3PO_4
1 buret clamp	
2 5-mL volumetric pipets	
1 2-mL graduated pipet (1/100)	
1 stirring plate	
1 micro magnetic stir bar	
5 25-mL volumetric flasks	
1 50-mL volumetric flask	
2 25-mL burets	

The first part of the experiment (**Part A**) measures the pH and alkalinity of the water samples and compares the values among them as well as with those of tap and distilled or DI water. The alkalinity will be measured by titration with dilute hydrochloric acid up to three specific pH values: 8.3, 7 (i.e., neutrality), and 4.5. This will make it possible to calculate the different contributions of the species responsible for the alkalinity. The acid–base indicator phenolphthalein is used to indicate the 8.3 endpoint and to obtain the P-alkalinity. Results can be best interpreted in the light of the discussion of Section 6.3.1.1 and Example 6.5.

To identify the amount of acid needed to reach the 4.5 endpoint, one may use methyl orange or bromocresol green indicators. This titration indicates the presence of the rest of the carbonate ions present (when it reaches a pH near 6.5) and of all the bicarbonate ions present up to pH 4.5. The total amount of acid indicates total alkalinity. For a more accurate titration, the pH must be followed with a pH meter. The student will compare the amount of each titrant added and the pH of each sample, besides determining the alkalinity and graphically determining the buffering capacity of each. The student will also identify the weak or strong buffers.

In the second part of the experiment (**Part B**) the student will prepare solutions with different proportions of carbonate and phosphate salts, and will titrate them to evaluate their buffering limits and the proportion that will yield the best buffering capacity. The student will observe the effect of varying the proportions of the acid to the conjugate base as well as of altering the concentration of the buffering constituents present. The differences among the conjugate systems used will also be observed.

In **Part C** the student will use the Henderson–Hasselbalch equation to determine the conjugate base/weak acid ratio to buffer a desired pH and prepare it experimentally. The initial pH and buffering capacity are then determined by titration, and they are monitored through the pH changes that result from adding specific amounts of standardized acid or base.

Safety Measures

The titrants should not come in contact with the skin or eyes because they may be corrosive. The student must consider all the safety measures normally taken when handling this kind of reactants. In case of spillage of an acid solution or of skin contact, wipe clean with a clean cloth and wash thoroughly and abundantly with water (sprinkle the table or surface with sodium bicarbonate). All of the residues generated in this experiment can be disposed of down the drain once they have been neutralized.

Experimental Sequence

Estimated time required: 5 min per sample

Part A. Measurement of alkalinity or ANC

Samples and procedure:

1. Collect samples (or prepare them synthetically) of: (1) river or lake water, (2) ocean water, (3) mineral water (or if available, use groundwater), (4) tap water, and (5) D.I. water. Use a calibrated

pH meter throughout the entire experiment and wait until stable readings are obtained.

2. Measure the pH and temperature of each sample. If the pH is above 8.3, determine the P-alkalinity. If it is not, only the M-alkalinity (i.e., methyl orange or bromocresol green alkalinity at pH $= 4.3$) can be measured.

3. If pH < 4.3, measure the acidity of the sample.

4. To prevent masking of the endpoint when using colored indicators, make sure the sample is colorless and free of turbidity. To have this condition, filter the sample prior to titration and—if the color were a problem—add a small amount of activated carbon (prior to filtration). Filtration can be done by means of a syringe equipped with a filter holder and a fine-pore filter. To eliminate any free chlorine that might interfere with the titration, add a drop or two of 0.01 M sodium thiosulfate.

5. Place a microburet in a stand, rinse it with a small amount of the 0.01 M acid titrant, and fill it to the desired mark.

6. With a volumetric pipet, take a 10-mL portion of the water sample and place it in a 25- or 50-mL beaker. Place a micro magnetic stir bar inside the beaker and put the beaker on a stirring plate. Immerse the bulb of the pH electrode in the liquid sample, without touching the stir bar. Measure the pH.

7. Add 2 to 3 drops of the phenolphthalein indicator (e.g., with a Beral pipet). Observe if any color appears.

8. If upon adding the P indicator there is no color, then the P-alkalinity is zero. In this case, immediately add 1 to 2 drops of the methyl orange (or of the bromocresol green) indicator and start titrating to the endpoint.

9. If a pink color appears upon adding the phenolphthalein indicator, titrate drop wise (in 0.1 or 0.2 mL increments), stir gently, and record the resulting pH and volume of titrant added. Note the pH reading at the point when the color of the indicator disappears (it must be close to 8.3). Then, immediately add 2 to 3 drops of the methyl orange (or the bromocresol green) indicator, and continue the titration until the exact 4.5 endpoint is reached (the solution turns salmon in the case of methyl orange or yellowish in color, in the case of the bromocresol green indicator).

10. Repeat the same technique for each water sample.

B. Buffering Capacity

Estimated time required: 15 min per sample

Method

B.1 Buffering capacity of natural samples. The measurements carried out in part A to determine the alkalinity of several water samples will also serve to determine the buffering capacity of each sample.

B.2 Factors that affect the pH and buffering capacity

B.2.1 Prepare a $Na_2CO_3 + NaHCO_3$ solution by introducing with a volumetric pipet 10 mL of freshly-prepared 0.1 M solutions of each salt into a 25-mL volumetric flask. Add D.I. water up to the mark. Mix perfectly and pour the resulting solution into a 50-mL beaker.

(a) Based on the known pKa values and the Henderson–Hasselbalch equation, calculate the pH of this solution.

(b) Immediately after preparing the solution, measure its pH by placing the pH electrode inside the solution and reading the stabilized value. Remember that if the solution is allowed to rest for a long time, atmospheric CO_2 will dissolve in it and the pH values will be altered.

(c) Rinse a 25-mL buret with 0.25 M HCl and fill it with the same acid solution. Start titrating the solution with the pH probe inside, gently mixing with the magnetic stirrer. Add small increments of acid. Record the volume of acid added and the resulting pH; make sure that by adding the acid there is a change of more than one pH unit in the resulting solution. Preferably in the last part of the titration, switch from the 25-mL buret to a microburet in order to add small increments (e.g., 0.1 mL) up to the lowest constant pH value attainable. Record all your data.

B.2.2 Proceed exactly as in point B.2.1, but use 5 mL of each carbonate and bicarbonate solution

TABLE C.1 K_a and pK_a Values for the Acid and Conjugate Base of the Phosphoric Acid System

Acid	Conjugate base	K_a	pK_a
H_3PO_4	$H_2PO_4^-$	7.62×10^{-3}	2.12
$H_2PO_4^-$	HPO_4^{2-}	6.23×10^{-8}	7.21
HPO_4^{2-}	PO_4^{3-}	2.2×10^{-13}	12.67

(instead of 10 mL) and follow the same technique as established in B.2.1 (a) through (c).

B.2.3 Proceed exactly as in point B.2.1, but use 10 mL of the Na_2CO_3 solution and 1 mL of the $NaHCO_3$ solution.

B.2.4 Proceed exactly as in point B.2.1, but use 10 mL of the $NaHCO_3$ solution and 1 mL of the Na_2CO_3 solution.

B.2.5 Proceed exactly as in all the previous steps, but use 5 mL of a KH_2PO_4 solution and 5 mL of a K_2HPO_4 solution.

C. Calculation, preparation and evaluation of a specified pH buffer solution

Estimated time required: 20 min per sample

Method

C.1 Based on the K_a values for the phosphoric system (see Table C.1) and the Henderson–Hasselbalch equation, the student groups will prepare 50-mL of a phosphate buffer with a requested pH. (The proportions used in experiment B.2.5 are not permitted here). For example, the groups can prepare buffers with pH values of 4, 5, 6, 7, 8, or higher.

Each group will then take its corresponding buffer and measure the acid buffering capacity and base buffering capacity, using the following method. Use a calibrated pH meter throughout the entire experiment and wait until stable readings are obtained.

C.2.a Immediately after preparing the phosphate buffer solution, place 25 mL in a 50-mL beaker and measure its pH.

C.2.b Proceed as in point B.2.1 (c).

C.2.c Repeat the process with another 25 mL aliquot of buffer solution, but this time titrate with 0.25 M NaOH. Record all your titration volumes.

Name_____Section_____Date_____

Instructor_____Partner_____

PRELABORATORY REPORT SHEET—EXPERIMENT 3

Experimental Title: _____

Objectives

Flow sheet of procedure

Waste containment procedure

PRELABORATORY QUESTIONS AND PROBLEMS

1. Explain why it is important to know the ANC of a water sample.
2. Explain the meaning and importance of the "buffering capacity of a natural water system."
3. Which are the main natural buffering systems and the predominating pH values of the main types of waters on Earth?
4. Explain what the *sensitivity to an acid discharge* for an aqueous natural system means and what parameters define this sensitivity.
5. Use the Henderson–Hasselbalch equation to explain under what conditions the $pH = pK_a$.
6. State how to calculate (and perform the calculation) the necessary amounts needed to prepare 500 mL of a phosphate buffer having a pH of_____ if you have 1 M solutions of each of the species of the phosphate system.
7. Explain what the buffering index is, and what parameters typically affect its value.
8. What is the difference between the acid buffering capacity and the base buffering capacity of a system?
9. Based on your response to question 6, calculate the buffering index for that specific buffer.
10. Explain how to determine experimentally the buffering index of a sample.
11. In which of the five kinds of water samples would you expect to find a higher buffering capacity? Why?
12. Explain what is a Gran plot titration for ANC calculation; how it is carried out, and what are the equations involved.

Additional Related Projects

- Use buffer systems with different components from those used here (e.g., citric acid, formic acid, ammonium chloride, ascorbic acid, and acetic acid) and follow the procedures described above.
- From the systems described in the previous related project, select the most adequate acid-conjugate base system and the acid/base proportion required in order to obtain a desired pH. Test your prediction experimentally.

Name_____Section_____Date_____

Instructor_____Partner_____

LABORATORY REPORT SHEET.
POSTLABORATORY PROBLEMS AND QUESTIONS—
EXPERIMENT 3

PART A. ALKALINITY or ANC

Origin of sample:

#1_____

#2_____

#3_____

Visible characteristics of the sample (color, odor, suspended solids present, etc.)

#1_____

#2_____

#3_____

Precautions observed during the sampling. Sampling procedure for each sample:

#1_____

_____

#2_____

_____

#3_____

_____

EXPERIMENTAL DATA

1) pH

pH of the samples:

#1_____ _____

#2_____ _____

#3_____ _____

#4 Tap water_____ _____

#5 D.I. water_____ _____

2) Alkalinity measurement:

SAMPLE #1:

Volume of water sample:_____

Titrant:_____Titrant concentration: _____

mL of titrant	pH	mL of titrant	pH

Volume of titrant to reach pH = 8.3:_____

Volume of titrant to reach pH = 4.5:_____

SAMPLE #2:

Volume of water sample: _____

Titrant: _____ Titrant concentration: _____

mL of titrant	pH	mL of titrant	pH

Volume of titrant to reach pH = 8.3:_____

Volume of titrant to reach pH = 4.5:_____

SAMPLE #3:

Volume of water sample: _____

Titrant: _____ Titrant concentration: _____

mL of titrant	pH	mL of titrant	pH

mL of titrant	pH	mL of titrant	pH

Volume of titrant to reach pH= 8.3:_____

Volume of titrant to reach pH= 4.5:_____

SAMPLE #4:

Volume of water sample: _____

Titrant: _____Titrant concentration: _____

mL of titrant	pH	mL of titrant	pH

Volume of titrant to reach pH = 8.3:_____

Volume of titrant to reach pH = 4.5:_____

SAMPLE #5:

Volume of water sample: _____

Titrant: _____ Titrant concentration: _____

mL of titrant	pH	mL of titrant	pH

Volume of titrant to reach pH = 8.3:_____

Volume of titrant to reach pH = 4.5:_____

PART A. Data analysis:

1. From the experimental data obtained with each sample tested, report the alkalinity values and the concentration of each alkalinity species present (as mg/L of $CaCO_3$).

Why is distilled water considered to have no buffering capacity?

_____

_____

To calculate the alkalinity, apply the following formula:

$$\text{Alkalinity, mg/L } CaCO_3 = \frac{(V \text{ acid, mL}) (M, \text{ acid solution})(100 \text{ g/mol } CaCO_3) (1000 \text{ mg/g})}{(V \text{ of sample, mL})}$$

Sample	P-Alkalinity as CaCO$_3$	M-Alkalinity as CaCO$_3$	Total Alkalinity as CaCO$_3$
#1			
#2			
#3			
#4			
#5			

To determine the approximate concentration of hydroxide, carbonate and bicarbonate ions in each sample, one can use the following relationships and assumptions:

Volume of titrant to reach the endpoint	Predominant chemical species	Concentration of the chemical species
$V_p = 0$	HCO$_3^-$	= M alkalinity
$V_M = 0$	OH$^-$	= P alkalinity
$V_P = V_M$	CO$_3^{2-}$	= P alkalinity
$V_P > V_M$	OH$^-$ CO$_3^{2-}$	= (P − M) alkalinity = M alkalinity
$V_p < V_M$ and the pH is between 8.2–9.6	CO$_3^{2-}$ HCO$_3^-$	= 2P alkalinity = (M − 2P) alkalinity
$V_P < V_M$ and pH > 9.6	CO$_3^{2-}$ OH$^-$	= 2(M − P) alkalinity = (2P − M) alkalinity

Note: The P (phenolphthalein) endpoint = pH 8.3, and the M (methyl orange or bromocresol green) endpoint = pH 4.5.

The dominant species at 4.5 are assumed to be bicarbonate and carbonate, and when OH$^-$ ions are present, no bicarbonate ions can be present. It is also assumed that [H$^+$] is not relevant in alkaline pH values, and that one half of the carbonate ions present become neutralized at the 8.3 endpoint.

With the above information, determine the speciation of the alkalinity in each sample:

Sample	$[H^+]$	$[OH^-]$	$[HCO_3^-]$	$[CO_3^{2-}]$
#1				
#2				
#3				
#4				
#5				

2. Observe the differences between the values of these samples and indicate what has influenced the composition of these samples.

#1 _____   _____

#2 _____   _____

#3 _____   _____

Tap water
#4 _____   _____

D.I. water
#5 _____   _____

3. Based on the experimental values of ANC obtained with the different samples, classify these samples according to their sensitivity to acid discharge.

Sample	Sensitivity	Justification
#1		
#2		
#3		
#4		
#5		

PART B. BUFFERING CAPACITY

B.1.a EXPERIMENTAL DATA

SAMPLE #1: _____

Volume of water sample: _____

Titrant: _____Titrant concentration:_____

mL of titrant	pH	ΔV_b	ΔpH
		-	-

Estimate the pH value at which the maximum rate of change of pH per unit of volume (i.e., $\Delta pH/\Delta V$) occurs: _____

SAMPLE #2: _____

Volume of water sample: _____

Titrant: _____ Titrant concentration: _____

mL of titrant	pH	ΔV_b	ΔpH
		-	-

Estimate the pH value at which the maximum rate of change of pH per unit of volume (i.e., $\Delta pH/\Delta V$) occurs: _____

SAMPLE #3: _____

Volume of water sample: _____

Titrant: _____ Titrant concentration:_____

mL of titrant	pH	ΔV_b	ΔpH
		-	-

Estimate the pH value at which the maximum rate of change of pH per unit of volume (i.e., ΔpH/ΔV) occurs: _____

SAMPLE #4: Tap water

Volume of water sample: _____

Titrant: _____Titrant concentration:_____

mL of titrant	pH	ΔV_b	ΔpH
		-	-
		-	-

Estimate the pH value at which the maximum rate of change of pH per unit of volume (i.e., $\Delta pH/\Delta V$) occurs: _____

B.1.b Determine the buffering capacity of each sample:

For this purpose, you must consider the largest volume of base added in order to raise the pH by one unit. Then, convert this volume into moles of base consumed per initial volume of buffer (considered in liters).

Sample	Buffering capacity
#1	
#2	
#3	
#4	
#5	

Which of the measured samples has the highest buffering capacity?

At what pH does this solution buffer?_____

B.2 Factors that affect the buffering capacity

B.2.A

Buffer B.2.1:

Buffer: _____

Volume of sample: _____

Titrant: _____ Titrant concentration:_____

mL of titrant	pH	ΔV_b	ΔpH
0		-	-

mL of titrant	pH	ΔV_b	ΔpH

Estimate the pH value at which the maximum rate of change of pH per unit of volume (i.e., $\Delta pH/\Delta V$) occurs: _____

Buffer B.2.2

Buffer: _____

Volume of sample: _____

Titrant: _____ Titrant concentration: _____

mL of titrant	pH	ΔV_b	ΔpH

Estimate the pH value at which the maximum rate of change of pH per unit of volume (i.e., ΔpH/ΔV) occurs: _____

Buffer B.2.3

Buffer: _____

Volume of sample: _____

Titrant: _____ Titrant concentration:_____

mL of titrant	pH	ΔV_b	ΔpH
		-	-

Estimate the pH value at which the maximum rate of change of pH per unit of volume (i.e., $\Delta pH/\Delta V$) occurs: _____

Buffer B.2.4

Buffer: _____

Volume of sample: _____

Titrant: _____ Titrant concentration: _____

mL of titrant	pH	ΔV_b	ΔpH
		-	-

Estimate the pH value at which the maximum rate of change of pH per unit of volume (i.e., $\Delta pH/\Delta V$) occurs: _____

Buffer B.2.5

Buffer: _____

Volume of sample: _____

Titrant: _____Titrant concentration:_____

mL of titrant	pH	ΔV_b	ΔpH
		-	-

Estimate the pH value at which the maximum rate of change of pH per unit of volume (i.e., $\Delta pH/\Delta V$) occurs: _____

B.2.b.1 Determine the buffering capacity of each sample:

For this purpose, you must consider the largest volume of base added in order to raise the pH by one unit. Then, convert this volume into moles of base consumed per initial volume of buffer (considered in liters).

Buffer	Buffering capacity
B.2.1	
B.2.2	
B.2.3	
B.2.4	
B.2.5	

Which of the measured samples has the highest buffering capacity?

At what pH does this solution buffer?_____

B.2.b.2 Determine theoretically the buffering index of each of the buffers mentioned above.

Buffer	Buffering Index
B.2.1	
B.2.2	
B.2.3	
B.2.4	
B.2.5	

B.2.b.3 Determine experimentally the buffering index of each of the buffers studied.

Buffer	Buffering Index
B.2.1	
B.2.2	
B.2.3	
B.2.4	
B.2.5	

Include the pertinent graphs and calculations.

B.2.b.4 Relate the differences in composition of the different buffers to the corresponding responses: pH, buffering capacity, buffering index. What do you conclude with respect to their differences and responses?

PART C: CALCULATION, PREPARATION, AND EVALUATION OF A SPECIFIED pH BUFFER SOLUTION

Buffer assigned to your team: _____

Include all the calculations needed to arrive at the composition and volumes used in the preparation of the corresponding buffer:

C.1.a Acid buffering capacity

Buffer: _____

Volume of sample: _____

Titrant: _____Titrant concentration:_____

mL of titrant	pH	ΔV_b	ΔpH

Estimate the pH value at which the maximum rate of change of pH per unit of volume (i.e. $\Delta pH/\Delta V$) occurs: _____

C.1.b Base buffering capacity

Buffer: _____

Volume of sample: _____

Titrant: _____Titrant concentration:_____

mL of titrant	pH	ΔV_{ac}	ΔpH

Estimate the pH value at which the maximum rate of change of pH per unit of volume (i.e., $\Delta pH/\Delta V$) occurs: _____

C.2.a How does the experimental pH value of the buffer solution that you prepared compare with the theoretical pH? Is there a deviation greater than 0.5 pH units? Why do you think this is so?

C.2.b Determine experimentally the acid and base buffering capacity of your buffer solution: For this purpose, you must consider the largest volume of base added in order to raise the pH by one unit. Then, convert this volume into moles of base consumed per initial volume of buffer (considered in liters).

	Buffering capacity
Acid	
Base	

Include all your calculations.

Considering the other buffers prepared in part B, how does yours compare in buffering capacity?

C.2.b.2 Determine theoretically the buffering index of the prepared buffers.

Buffer	Buffering Index

Include all the equations and calculations.

C.2.b.3 Determine experimentally the buffering index of the prepared buffer.

Buffer	Buffering Index

Include all the pertinent graphs and calculations.

Was your prediction about the buffering index of your assigned solution correct?
_____

What percentage deviation did it present? _____

*Suspected reasons:*_____

If the solution is a buffer, in what pH range does it buffer? _____

Student Comments and Suggestions

Literature References

Andersen. C. B. "Understanding Carbonate Equilibria by Measuring Alkalinity in Experimental and Natural Systems," *J. Geochem. Educ.* **2002**, *50*, 389–403.

Dunnivant, F. M. Experiment 21 (Determination of Alkalinity of Natural Waters) in: *Environmental Laboratory Exercises for Instrumental Analysis and Environmental Chemistry*; Wiley-Interscience: New York, 2004.

Mihok, M. Keiser, J. T.; Bortiatynski, J. M.; Mallouk, T. E. Experiment 2 (Iron and Alkalinity Determinations) in: "An Environmentally-Focused General Chemistry Laboratory", *J. Chem. Educ.* **2006**, *83*, 250. http://jchemed.chem.wisc.edu/Journal/Issues/2006/FebACS/ACSSub/ACSSupp/JCE2006p0250W.pdf

University of Wisconsin-Stout, Chemistry Department. Experiment 8 (Acidity and Alkalinity of Drinking Water) in: *Environmental Chemistry Lab Manual. Laboratory and Lecture Deconstrations.* http://www.uwstout.edu/faculty/ ondrusm/ondrusm/manual/index.html

Experiment 4
Aqueous Carbonate Equilibria and Water Corrosiveness

Reference Chapters: 2, 5, 6

Objectives

After performing this experiment, the student shall be able to:

- Determine experimentally if a calcium carbonate dissolution reaction has reached equilibrium.
- Measure equilibrium-determining parameters and use them to predict the corrosive or deposit-forming capacity of an aqueous solution by applying Langelier's and Ryznar's indexes.

Introduction

Natural aqueous carbonate equilibria play key roles in the characteristics of a water body or a water sample and its buffering capacity. In Nature, such equilibria depend on the solubility constant and therefore on the concentration of Ca^{2+} and the various carbonate forms. These equilibria will be highly dependent on pH, CO_2 concentration in air, CO_2 solubility, temperature, and pressure. Carbon dioxide in air dissolves in water, and this process is governed by Henry's constant:

$$CO_{2(g)} \rightleftarrows CO_{2(aq)} \qquad (1)$$

The carbonate equilibrium is defined by the following equations:

$$CO_{2(aq)} + H_2O \rightleftarrows H_2CO_3 \quad K_{eq} = 3.5 \times 10^{-2} \quad (2)$$

$$H_2CO_3 \rightleftarrows H^+ + HCO_3^- \quad K_{a1} = 4.2 \times 10^{-7} \quad (3)$$

$$HCO_3^- \rightleftarrows H^+ + CO_3^{2-} \quad K_{a2} = 4.8 \times 10^{-11} \quad (4)$$

It is clear from these equations that the concentration of each carbonate species is a function of pH.

In circumneutral values of pH, $CaCO_3$ is rather insoluble, as can be deduced from eq. 5:

$$CaCO_{3(s)} \text{ (calcite)} \rightleftarrows Ca^{2+} + CO_3^{2-}$$
$$K_{sp} = 4.8 \times 10^{-9} \qquad (5)$$

(Values of constants at 25°C).

In circumneutral solutions, the second dissociation of H_2CO_3 (eq. 4) is negligible, and the H^+ ions from the first dissociation (eq. 3) react with $CaCO_3$ as follows:

$$CaCO_{3(s)} + H^+ \rightleftarrows Ca^{2+} + HCO_3^- \qquad (6a)$$

and therefore, the overall dissolution reaction of calcite in the presence of aqueous CO_2 is:

$$CaCO_{3(s)} + H_2O_{(1)} + CO_{2(aq)} \rightleftarrows Ca^{2+} + 2HCO_3^- \qquad (6b)$$

As discussed in Chapter 2, if one considers an ideal (i.e., infinitely dilute) solution, an equilibrium equation for reaction 6a (based on the activity of the species and given by K'_{eq}) can be approximated by using the concentrations as follows:

$$K_{eq} = \{[Ca^{2+}][HCO_3^-]\}/[H^+] \qquad (7)$$

The consequence of these reactions is that the water can be either aggressive (i.e., corrosive), or precipitating (i.e., tends to form $CaCO_3$ deposits). Any process that increases the amount of CO_2 present will increase the H^+ concentration and displace the reaction toward the dissolution of calcite. The opposite process will favor the formation of calcite deposits. The first case may generate pitting in water distribution systems, and the latter may plug up water pipes with deposits (especially if the water is heated, since $CaCO_{3(s)}$ is one of the few compounds that display inverse solubility).

In an open system (as in a lake), $CaCO_{3(s)}$ and H_2O will be in contact with atmospheric CO_2 and thus will tend to achieve the above equilibria. The presence of calcareous material in lake beds increases the buffering capacity of the lake.

In closed systems there is less contact with atmospheric CO_2 as well as a smaller possibility for its volatilization. This is the case of groundwater, where only small amounts of CO_2 would be present derived from low biological activity in the surrounding soil due to its poor organic content. This affects the pH of water and therefore the concentration of each ion, as predicted by the equations just discussed.

To find out whether reaction 6 has reached equilibrium, one can determine its free energy change. If it is positive, the reaction will tend towards precipitation, and if it is negative, the dissolution of calcite will dominate. This free energy change is given by:

$$\Delta G = \Delta G° + RT \ln Q' \qquad (8)$$

where

$$\Delta G^0 = -RT \ln K'_{eq} = -RT \ln [\{(\gamma_{Ca^{2+}[Ca^{2+}]_{eq}} \cdot \\ \gamma_{HCO_3^-[HCO_3^-]_{eq}})/\gamma_{H^+[H^+]_{eq}}\}] \qquad (9)$$

and

$$Q' = \text{conditional reaction quotient}$$
$$= (\gamma_{Ca^{2+}[Ca^{2+}]} \cdot \gamma_{HCO_3^-[HCO_3^-]})/ \\ (\gamma_{H^+[H^+]}) \qquad (10)$$

If one can measure the concentrations of the species involved, the equilibrium character of the water (i.e., corrosive or precipitating) can be determined.

Another approach is to measure the ion product of $[Ca^{2+}]$ and $[CO_3^{2-}]$, and compare it to the solubility product of $CaCO_{3(s)}$, considering the concentrations as an approximation of each species activity or correcting it with the activity coefficient:

$$[Ca^{2+}][CO_3^{2-}] = K_{sp} \qquad (11)$$

This is also known as the *driving force index, DFI* defined as:

$$DFI = \{[Ca^{2+}][CO_3^{2-}]\}/K_{sp} \qquad (12)$$

The closer the *DFI* is to unity, the more stable will the water be; as it deviates from that value, water will be corrosive or deposit-forming.

Another method of measuring the departure from equilibrium is to compare the ion product through several practical indexes that are strongly related to corrosion or to deposit-forming tendencies of the

solution. The most common indexes are Langelier's and Ryznar's.

In both cases it is important to know the value of pH at which water with a given $[Ca^{2+}]$ and alkalinity is at saturation equilibrium (at a given temperature). Then, this value is compared to the actual pH of the solution. The formulas for such indexes are given below.

(a) *Langelier index, LI*

$$LI = pH - pH_s \qquad (13)$$

where pH_s is the pH required for saturation. The different tendencies of the solution under analysis are given in the following scheme.

	Solution	Tendency
LI		
	Saturated	To precipitate
0	At equilibrium	At equilibrium with the solid
	Unsaturated	To dissolve the solid further

The condition of $LI = 0$ is seldom observed in waters used in industry since it is preferable to promote the formation of a slight deposit on iron pipes for protective reasons.

A disadvantage of the LI is that it does not consider the calcium complexes formed at pH > 8. In fact, the LI is only valid between pH 6 and 8.

(b) *Ryznar index, RI*

This empirical index is calculated with the following equation:

$$RI = 2pH_s - pH \qquad (14)$$

RI	Tendency of the water
	Very corrosive and very acidic
8.5	
8.0	
7.5	Corrosive and acidic
7.0	
6.5	At equilibrium and essentially neutral
6.0	
5.5	Tends to form deposits or precipitates
	Forms deposits or precipitates easily

and the tendency of the water is given in the following scheme.

To calculate pH_s, one determines the pH of the equilibrium reaction from the relationships in equations 4, 5 and 15:

$$pH_s = -\log[H^+]_s \qquad (15)$$

where $[H^+]_s$ is the hydrogen ion concentration at saturation. Then,

$$pH_s = -\log[H^+]_s = -\log\frac{K_{a_2}[Ca^{2+}][HCO_3^-]}{K_{sp}} \qquad (16)$$

where

$$K_{a2} = \frac{[CO_3^{2-}][H^+]}{[HCO_3^-]}$$

and K_{sp} is taken from eq. 11. Then,

$$pH_s = -\log(K_{a_2}/K_{sp}) - \log[Ca^{2+}] - \log[HCO_3^-] \qquad (17)$$

or

$$pH_s = -\log(K_{a_2}/K_{sp}) + p[Ca^{2+}] + p[HCO_3^-] \qquad (18)$$

From the simplified definition of alkalinity (i.e., considering only the hydrogen, hydroxide, bicarbonate, and carbonate ions contribution to the alkalinity; see Section 6.3) one has:

$$Alkalinity = [HCO_3^-] + 2[CO_3^{2-}] + [OH^-] - [H^+] \qquad (19)$$

As stated above, the LI is valid only at $6 < pH < 8$, where the predominant carbonate form in natural waters is the bicarbonate ion. Here, $Alk \gg [H^+]$, $[OH^-]$. Then, $ALK \approx [HCO_3^-]$ and therefore the equation for pH_s can be rewritten as:

$$pH_s = pK_{a_2} - pK_{sp} + p[Ca^{2+}] + p[Alk] \qquad (20)$$

If a more rigorous calculation is desired for cases where the solutions have higher ionic concentrations, the chemical equilibrium concentrations will be significantly affected by the activity coefficients. Therefore, equation 20 will be expressed as follows:

$$pH_s = pK_{a_2} - pK_{sp} + p[Ca^{2+}] + p[Alk]$$
$$- \log \gamma_{Ca^{2+}} - \log \gamma_{HCO_3^-} \qquad (21)$$

where $\gamma_{Ca^{2+}}$, $\gamma_{HCO_3^-}$ are the activity coefficients of the corresponding ions and depend on the ionic strength, I, of the solution. Therefore, these can be calculated

with the following approximation by Guntelberg of the extended Debye–Hückel equation.

$$\log \gamma_i = [0.5(Z_i^2)I^{0.5}]/[1 + I^{0.5}] \qquad (22)$$

where Z_i is the charge of each ith ion present, and I is expressed as a function of the total dissolved solids, TDS (in mg/L):

$$I = 2.5 \times 10^{-5} \text{ TDS} \qquad (23)$$

Another rigorous approach for the calculation of pH_s, involves making explicit the alkalinity equation as a function of $[Ca^{2+}]$ and $[H^+]_s$. Then, the other concentrations are substituted by their equivalents as follows:

$$Alkalinity = [HCO_3^-] + 2[CO_3^{2-}] + [OH^-] - [H^+] \qquad (24)$$

$$[OH^-] = K_w/[H^+]_s \qquad (25)$$
$$[CO_3^{2-}] = K_{sp}/[Ca^{2+}] \qquad (26)$$
$$[HCO_3^-] = [H^+][CO_3^{2-}]/K_{a_2} \qquad (27)$$
$$[HCO_3^-] = \{[H^+]K_{sp}\}/\{K_{a_2}[Ca^{2+}]\} \qquad (28)$$

Therefore:

$$Alkalinity = \{[H^+]_s K_{sp}\}/\{K_{a_2}[Ca^{2+}]\} + 2(K_{sp}/[Ca^{2+}])$$
$$+ (K_w/[H^+]_s - [H^+]_s) \qquad (29)$$

The value of $[H^+]_s$ is solved with equations 20 and 29 either by an iterative process, a regression, or a numerical method. Then this value is corrected with its corresponding activity coefficient as a function of the total dissolved solids, using equations 22 and 23.

To correct the equilibrium constants for temperature (in degrees K), one can use the standard equations proposed for the range of 273 to 373 K.

$$pK_w = (4471/T) + 0.01706 \, T - 6.087 \qquad (30)$$
$$pK_{a_2} = 107.88 + 0.0325 \, T - (5151.79/T)$$
$$- 38.926 \log T + (563713.9/T^2) \qquad (31)$$
$$pK_{sp} = 171.907 + 0.0078 \, T$$
$$- (2839.32/T) - 71.595 \log T \qquad (32)$$

The method used in the present experiment consists in analyzing for pH, Alk, $[Ca^{2+}]$, and TDS in a series of synthetic or natural water samples, in order to determine their characteristics with respect to the calcite dissolution equilibrium and to the corrosive or deposit-forming potential.

Experimental Procedure

Estimated time to complete the experiment: 3 h (approx. 15 minutes per sample + 5 minutes for weighing before and after the sample is dried in the oven).

Materials	Reagents
1 pH meter	0.002 M H_2SO_4
5 500-mL beakers	0.01 M H_2SO_4
2 magnetic stirrers	phenolphthalein indicador
6 small stir bars	(0.1 g of phenolphthalein in
0.8 micron acetate filter	100 mL of a 1:1 water: ethanol
membranes	solution)
1 25-mm plastic filter	bromocresol green indicator or a
holder	mixed indicator (e.g., 0.02 g of
1 10-mL syringe	methyl red and 0.1 g of
1 propipet	bromocresol green in 100 mL
2 5-mL graduated pipet	of ethanol)
2 5-mL volumetric pipet	0.01 M NaOH
12 10-mL beakers	standard buffer solutions
12 25-mL beakers	(pH 4, 7 and 10)
1 analytical weighing	murexide indicator
balance	0.001 M EDTA
1 drying oven (105°C)	D.I. water
1 dessicator	$CaCO_3$
1 crucible tong	$NaHCO_3$
2 crystallizing dishes	$Ca(OH)_2$
(medium size)	
1 microburet	
(see Experiment 1)	
1 2-mL volumetric pipet	
1 spatula	
2 Beral pipets	
1 thermometer	
1 2-mL graduated pipet	
1 25-mL graduated cylinder	
6 50-mL Erlenmeyer flasks	
3 25-mL Erlenmeyer flasks	
1 universal stand	
1 three-finger buret clamp	

Safety Measures

One must be careful with the titrant for the alkalinity test so as to prevent it from coming into contact with the skin or eyes, since sulfuric acid is corrosive. All the residues generated in this experiment can be disposed of down the drain once they have been neutralized.

For each sample, the following techniques must be followed:

(a) pH measurement
(b) alkalinity measurement
(c) calcium ion concentration measurement
(d) total dissolved solids

Note: Some of these techniques have already been discussed in experiments 1–3.

Method

A. Effect of pH on the characteristics of a calcite solution sample

Use D.I. water to prepare 500 mL of a $CaCO_3$ solution containing 0.5 g (do this 1 week in advance in an open vessel). This calcite solution is called *solution A*. Measure its pH and temperature. Place a 25-mL portion in a graduated cylinder and then place it in a 50-mL Erlenmeyer flask. Repeat the operation preparing five more flasks. Number each sample (including the first one described above) as: 1, 2, 3, 4, 5, and 6. To the first three flasks add increasing amounts of 0.01 M H_2SO_4 (1, 2, and 4 mL, respectively), and to the other three flasks add 0.01 M NaOH (1, 3, and 5 mL, respectively). Mix each sample thoroughly (most contain solids), measure their pH, and then allow them to react by mixing with the magnetic stirrers or a rotary shaker for at least 2 h. After this time, perform the measurements for each sample established in part B including a sample of the original A solution.

B. Characterization of each water sample

Prepare the following:

- Solution B: Add 0.9 g of $CaCO_3$ to carbonate-free D.I. water (previously boiled and capped D.I. water) for a total of 500 mL. Prepare this solution in a closed vessel, without any free space, **one week** prior to the experimental session.
- Solution C: Add 0.9 g of $CaCO_3$ to D.I. water for a total of 500 mL. Prepare this solution in an open vessel, **one day** prior to the experimental session.
- Solution D: Add 0.9 g of $CaCO_3$ to D.I. water for a total of 500 mL. Prepare this solution in an open vessel, **the same day** of the experimental session.
- Solution E: Prepare a 0.1 M calcium bicarbonate solution by reacting 0.05 moles of $Ca(OH)_2$ and 0.1 moles of $NaHCO_3$ in 500 mL of D.I. water.
- Solution F: Use a natural water sample (e.g., groundwater).

Then, proceed as follows:

1. Measure the temperature and pH of each water sample (solutions A through F and 1 through 6, after the 2 h of reaction). Then carry out each of the following determinations with each solution.
2. Filter approximately 15 mL of the sample into a beaker, and take a 2-mL portion with a volumetric pipette. Place it in a 25-mL Erlenmeyer flask and titrate for alkalinity with dilute H_2SO_4 solution using phenolphthalein indicator, then after the endpoint, add the bromocresol indicator with a Beral pipet and titrate to the subsequent endpoint. The total volume of titrant will equal the total alkalinity. Note each of the volumes used in the titration.
3. Take another filtered 2 mL sample and titrate for calcium, first adding 0.5 mL of 0.1 M NaOH until basic and then adding a few crystals of murexide indicator or other calcium indicator. Titrate with 0.001 M EDTA to the endpoint. Note the total volume of titrant used. Repeat the process with each problem solution.
4. Weigh a marked (with a pen) previously oven dried 10-mL beaker and note the value. Now take a filtered sample with a 5 mL volumetric pipette and place it in one of the beakers. Repeat the process with each problem solution. Place the beaker or beakers on a crystallizer in the oven and let them dry at 105°C for at least 3 h or more. After all the samples are dry, place them in a desiccator until they are at ambient temperature and then weigh them again. Note the values. The difference in weights will correspond to the total dissolved solids of the sample.

With the values obtained for each sample, the student will be able to determine how far each solution is from the equilibrium as well as its corrosive or deposit-forming potential using different indexes.

Name_____Section_____Date_____

Instructor_____Partner_____

PRELABORATORY REPORT SHEET—EXPERIMENT 4

Objectives

Flow sheet of procedure

Waste containment procedure

PRELABORATORY QUESTIONS AND PROBLEMS

1. Explain why natural carbonate equilibria are environmentally relevant.
2. Under what conditions will a calcium solution tend to precipitate as calcium carbonate?
3. Calculate and plot the chemical species distribution diagram for carbonate. In what pH ranges does each one of the three carbonate species predominate?
4. What is the main difference between an open and a closed system for carbonate equilibria?
5. Explain the dependence of the dissolved concentration of CO_2 in water with respect to its corrosive or non-corrosive properties.
6. What other parameters (different from the above) contribute to the corrosive or non-corrosive nature of water?
7. What is the ionic strength of a solution? Why is the activity coefficient of a species in solution related to it?

8. What other parameters—besides the TDS—can be used to determine the ionic strength of a solution?
9. Explain the difference between a reaction quotient and the corresponding equilibrium constant.
10. Establish the electroneutrality or charge balance expression for calcium carbonate in water in contact with atmospheric CO_2. Develop this equation as an expression of $[H^+]$, p_{CO_2} (i.e., the partial pressure of carbon dioxide), Henry's constant and the corresponding equilibrium constants.

Additional Related Project

- Take an aliquot of one or more of the solutions prepared in this experiment, bring it to boil and repeat the evaluation of the filtered samples. This will demonstrate the effect of temperature.

Name_____Section_____Date_____

Instructor_____Partner_____

LABORATORY REPORT SHEET—EXPERIMENT 4

Part A. Equilibrium characteristics of different water samples (Corresponds to results from Part B of experimental procedure)

Experimental data.

1) pH measurement

WATER SAMPLE	pH	Temperature, °C
Sample A (prepared_____ days ago)		
Sample B (prepared_____ days ago; closed system)		
Sample C (prepared_____ h ago)		
Sample D (prepared_____ days ago)		
Sample E (saturated calcium bicarbonate solution)		
Sample F (source: _____)		

2) Alkalinity measurement:

Titrant: _____
Titrant concentration: _____
Sample volume: _____
Indicators: _____ *and*_____

WATER SAMPLE	mL of titrant for P-alkalinity	P-alkalinity, mg/L $CaCO_3$	mL of titrant for total alkalinity	Total alkalinity, mg/L $CaCO_3$	Total alkalinity, mol/L
Sample A					
Sample B					
Sample C					
Sample D					
Sample E					
Sample F					

Using the common equations based on the titration volumes, calculate:

WATER SAMPLE	$[OH^-]$, mol/L	$[CO_3^{2-}]$, mol/L	$[HCO_3^-]$, mol/L	$[H]^+$, mol/L
Sample A				
Sample B				
Sample C				
Sample D				
Sample E				
Sample F				

Include all equations and an example of your calculations.

3) Calcium concentration

Titrant: _____
Titrant concentration: _____
Indicators: _____
Sample volume: _____

WATER Sample	mL of titrant	$[Ca^{2+}]$ concentration, mg/L $CaCO_3$	$[Ca^{2+}]$ conc. mol/L
Sample A			
Sample B			
Sample C			
Sample D			
Sample E			
Sample F			

Equation and sample calculation:

4) Total dissolved solids
Sample volume: _____ mL

WATER SAMPLE	Weight of dry empty beaker, g	Weight of beaker with dried sample, g	Total dissolved solids in sample, mg	Total dissolved solids (TDS), mg/L
Sample A				
Sample B				
Sample C				
Sample D				
Sample E				
Sample F				

Sample calculation

Part B. Effect of pH on the solubility and equilibrium of calcium carbonate

Experimental data

1) pH and temperature measurement

WATER SAMPLE	Reactant added	mL added	pH	Temp., K
Treated sample 1				
Treated sample 2				
Treated sample 3				
Treated sample 4				
Treated sample 5				
Treated sample 6				

2) Alkalinity measurement:

Titrant: _____

Titrant concentration: _____

Sample volume: _____

Indicators: _____ and _____

WATER SAMPLE	mL of titrant for P-alkalinity	P-alkalinity, mg/L CaCO₃	mL of titrant for total alkalinity	Total alkalinity mg/L CaCO₃	Total alkalinity mol/L
Treated sample 1					
Treated sample 2					
Treated sample 3					
Treated sample 4					
Treated sample 5					
Treated sample 6					

Using the common equations based on the titration volumes, calculate:

WATER SAMPLE	$[OH^-]$, mol/L	$[CO_3^{2-}]$, mol/L	$[HCO_3^-]$, mol/L	$[H]^+$, mol/L
Treated sample 1				
Treated sample 2				
Treated sample 3				
Treated sample 4				
Treated sample 5				
Treated sample 6				

Include all equations and an example of calculations.

3) Calcium concentration

Titrant: _____

Titrant concentration: _____

Indicator: _____

Sample volume: _____

WATER SAMPLE	mL of titrant	$[Ca^{2+}]$ conc., mg/L $CaCO_3$	$[Ca^{2+}]$ conc., mol/L
Treated sample 1			
Treated sample 2			
Treated sample 3			
Treated sample 4			
Treated sample 5			
Treated sample 6			

   Equation and sample calculation:

4) Total dissolved solids

Sample volume: _____*mL*

WATER SAMPLE	Weight of dry empty beaker, g	Weight of beaker with dried sample, g	Total dissolved solids in sample, mg	Total dissolved solids (TDS), mg/L
Treated sample 1				
Treated sample 2				
Treated sample 3				
Treated sample 4				
Treated sample 5				
Treated sample 6				

   Example of calculation:

POSTLABORATORY RESULTS AND DISCUSSION—PART A

For each sample calculate the pHs and the other parameters requested.

WATER SAMPLE	Ionic strength	$log\gamma_{Ca^{2+}}$	$log\gamma_{HCO_3^-}$	Temp., K	pKa_2	pK_{sp}
Sample A						
Sample B						
Sample C						
Sample D						
Sample E						
Sample F						

WATER SAMPLE	pCa^{2+}	pAlk	pH_s	pH	Langelier Index	Ryznar Index
Sample A						
Sample B						
Sample C						
Sample D						
Sample E						
Sample F						

WATER SAMPLE	Reaction quotient, Q	$\Delta G°$	ΔG	Driving Force Index, DFI
Sample A				
Sample B				
Sample C				
Sample D				
Sample E				
Sample F				

Based on the above results, state your conclusions with respect to the characteristics of the solutions analyzed and classify each one. Justify each answer, on the basis of the values calculated and those measured.

WATER SAMPLE	Has the solution reached equilibrium?	What is the tendency of the solution?	Classification according to Langelier Index	Classification according to Ryznar Index
Sample A				
Sample B				
Sample C				
Sample D				
Sample E				
Sample F				

What are the differences between solutions A and B? What causes them?

What differences are observed among solutions A, C, and D? Is there a tendency with respect to the time elapsed since preparation?

What differences are noted between solutions A and E? What may have caused these differences?

According to your results, can the water source of Sample F be considered acceptable? Why?

How would the values change if the Langelier simplification of considering $[Alk] = [HCO_3^-]$ were substituted for the true calculated value? How much would it affect the result? Which result is more reliable?

POSTLABORATORY CALCULATIONS AND DISCUSSION— PART B

For each sample calculate the pHs and the other parameters requested. Do not consider here the value of alkalinity as equivalent to that of bicarbonates.

WATER SAMPLE	Ionic strength	$log\gamma_{Ca^{2+}}$	$log\gamma_{HCO_3^-}$	Temp., K	pKa_2	pK_{sp}
Treated sample 1						
Treated sample 2						
Treated sample 3						
Treated sample 4						
Treated sample 5						
Treated sample 6						

WATER SAMPLE	pCa^{2+}	$pHCO_3^-$	pH_s	pH	Langelier Index	Ryznar Index
Treated sample 1						
Treated sample 2						
Treated sample 3						
Treated sample 4						
Treated sample 5						
Treated sample 6						

WATER SAMPLE	Reaction quotient, Q	ΔG^o	ΔG	pH
Treated sample 1				
Treated sample 2				
Treated sample 3				
Treated sample 4				
Treated sample 5				
Treated sample 6				

Based on the above results, state your conclusions with respect to the characteristics of the solutions analyzed and classify each one. Justify each answer, on the basis of the values calculated and those measured.

WATER SAMPLE	Has the solution reached equilibrium?	What is the tendency of the solution?	Classification according to Langelier Index	Classification according to Ryznar Index
Treated sample 1				
Treated sample 2				
Treated sample 3				
Treated sample 4				
Treated sample 5				
Treated sample 6				

Conclusions:

Plot the pH_s vs pH values for the six treated samples + the original sample A. As a reference, draw a 45° line.

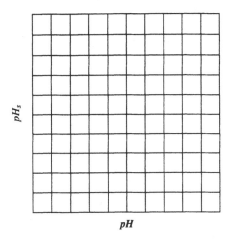

pH_s

pH

Observe the tendency and determine at what pH the solution would reach equilibrium.

Considering the amount of acid and base added, determine the amount of the corresponding substance required to reach equilibrium. For that purpose, draw a graph of mL of acid and base added vs. pH.

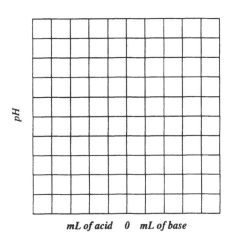

pH

mL of acid   0   mL of base

General conclusions:

Student Comments and Suggestions

Literature References

APHA/AWWA, *Standard Methods for the Examination of Water and Wastewater* , 18th ed.; Washington, 1992.

Sawyer, C. N.; McCarty, P. L.; Parkin, G. F. *Chemistry for Environmental Engineering* , 5th ed.; McGraw Hill: New York, 2003.

Stumm, W.; Morgan, J. J., *Aquatic Chemistry: Chemical Equilibria and Rates in Natural Waters*; 3rd ed.; Wiley Interscience: New York, 1996.

Vanderpool, D. "The pH Values for Cooling Water Systems", *The Analyst* **2004**, *11* (2) Spring issue. http://www.awt.org/members/publications/analyst/2004/spring/the_ph_values.htm

Experiment 5
The Point of Zero Charge of Oxides

Reference Chapters: 5, 6

Objectives

After performing this experiment, the student shall be able to:

- Understand the concept of point of zero charge (pzc) of a metal oxide in contact with water in an operational way.
- Estimate the pzc of a simple metal oxide by performing potentiometric titrations.
- Visualize the importance and applications of the pzc.

Introduction

Solid metal oxides in contact with aqueous solutions typically become hydrated and form a monolayer of surface hydroxyl groups that may become protonated or deprotonated, depending on the $H^+_{(aq)}$ concentration. This amphoteric behavior allows the oxide particles to develop electrical charges that are either positive (when they *receive* protons) or negative (when they *release* protons). In the following discussion we assume for simplicity that the surface species of such simple, hydrated metal oxides can be represented as MOH, and that the processes just described can be depicted as:

$$MOH_{(surf)} + H^+_{(aq)} \rightleftarrows MOH^+_{2\,(surf)} \quad (1)$$

$$MOH_{(surf)} + OH^-_{(aq)} \rightleftarrows MO^-_{(surf)} + H_2O_{(1)} \quad (2)$$

Alternatively one can write these reactions as deprotonations,

$$MOH^+_{2\,(surf)} \rightleftarrows MOH_{(surf)} + H^+_{(aq)} \quad (3)$$

$$MOH_{(surf)} \rightleftarrows MO^-_{(surf)} + H^+_{(aq)} \quad (4)$$

in which case the corresponding surface acidity constants (K_{a1}, K_{a2}) can easily be defined:

$$K_{a1} = (\{MOH\}[H^+])/\{MOH^+_2\} \quad (5)$$

$$K_{a2} = (\{MO^-\}[H^+]/\{MOH\} \quad (6)$$

where the braces indicate surface concentrations in mol/g of solid.

If the concentrations of the two types of sites resulting in equations 1 and 2 are the same, there will be no net charge on the surface. This condition is called the *point of zero proton charge, pzpc* (or *zero proton condition, zpc*). At solution pH values lower than that required for attaining the pzpc, the sites become protonated and an excess positive charge develops on the surface; here, the oxide behaves as a Brönsted acid and as an anion exchanger. The contrary occurs at pH values higher than the pzpc, where the oxide behaves as a Brönsted base and cation exchanger. Mixed oxides can have both exchange types, depending on the relative pK values of the different surface sites.

The pzpc can thus be measured by potentiometric titration if H^+ and OH^- are the only aqueous ions involved. In practice, however, electrolytes typically contain other anions, A^- and cations, C^+ that may adsorb onto the surface sites as follows:

$$MOH^+_{2\,(surf)} + A^-_{(aq)} \rightleftarrows MOH^+_2 A^-_{(surf)} \quad (7)$$

and/or

$$MOH^+_{2\,(surf)} + C^+_{(aq)} \rightleftarrows MO^- C^+_{(surf)} + 2H^+_{(aq)} \quad (8)$$

In these cases the net surface charge not only depends on the H^+ and OH^- ions in the medium, but also on the concentration of the electrolyte (that provides $A^-_{(aq)}$, $C^+_{(aq)}$). The pH of the aqueous solution at

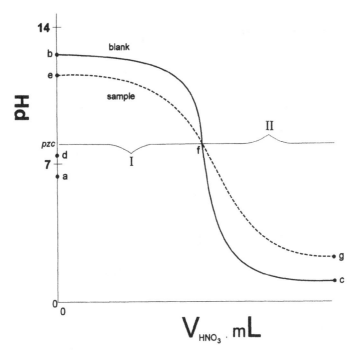

FIGURE 1. Determination of the point of zero charge, pzc of a simple metal oxide by a potentiometric mass titration technique, PMT. (Note: In this figure we assumed that pzc > 7, but the arguments apply to any other case as well, see Balderas, 2006 and Vakros, 2002). Reprinted from the Chemical Educator, vol. 11, No. 4, 2006, pp. 267–270, with permission.

which the sum of the all the surface positive charges balances the sum of all the surface negative charges is called the *point of zero charge, pzc*. In the absence of sorbed ions other than protons, the pzpc is equal to the pzc. (Note that the IUPAC recommends using periods after each letter in these abbreviations, but for the sake of simplicity we have omitted them).

A simple and fast method that gives a reasonable approximation to the pzc consists in locating the intersection of one or more titration curves performed with constant ionic strength and different amounts of solid, and that of a blank solution (i.e., without the solid). This is called the *potentiometric mass titration method, PMT*. This method is better understood by analyzing the plot in Figure 1 with the aid of the explanation given below.

The points *a*, *b*, and *c* in Figure 1 correspond to the blank solution, and *d*, *e*, *f*, and *g* correspond to the sample. The blank solution consists of *w* mL of a solution of an inert, supporting electrolyte (e.g., KNO_3) of concentration *x;* thus, the point at *a* corresponds to the natural pH of this solution. Upon addition of *y* mL of NaOH of concentration *z*, the pH of

the blank increases to *b*. By doing a potentiometric titration with a strong acid (e.g., HNO_3), the blank curve is generated (*c* is an arbitrary final point).

The initial pH of the sample solution (composed of *w* mL of the inert electrolyte solution plus *m* grams of the solid hydrated oxide, MOH) is at *d* (i.e., below the pzc due to the removal of H^+ from the solution by the neutral surface sites in MOH to form MOH_2^+ (eq. 1). Then, addition of the same *y* mL of *z* M NaOH to this suspension of the sample brings it to the point *e*, which is below *b*. This is because the OH^- added play three roles: they serve to neutralize the $H_{(aq)}^+$ and the $MOH_{2\,(surf)}^+$ created during the *a* → *d* step, and to remove $H_{(surf)}^+$ from the resulting MOH sites so as to create basic sites (MO^-) at the oxide surface. By doing a potentiometric titration, the $H_{(aq)}^+$ added from *e* to *f* (see Region I of the sample curve) play two roles: some of them neutralize the $OH_{(aq)}^-$, and the others neutralize the $MO_{(surf)}^-$. This is the reason for the smaller pH values in this curve as compared to those in the blank curve.

On the other hand, the $H_{(aq)}^+$ added from *f* to *g* (Region II of the sample curve) also play two roles:

a minor role, i.e., to neutralize the (small amount of) $OH^-_{(aq)}$ remaining in solution and a major role, i.e., to protonate the neutral $MOH_{(surf)}$ sites. The difference in pH between the sample and blank curves in this region will therefore be essentially proportional to the amount of solid sample in the suspension.

Between the end of Region I and the beginning of Region II in the sample curve, there is a small region (or even a single point, f) where all the $H^+_{(aq)}$ added serve to neutralize the $OH^-_{(aq)}$, just as they do in the blank curve. For this reason both curves intersect at this point, which is then identified as the pzc. This intersection should appear at the same pH value for any amount of oxide placed in the solution.

Literature results also show that a *single* determination curve intersecting the blank curve gives a very good approximation to the true value (see Vakros, 2002). (We hereby call this the *simplified potentiometric mass titration method, sPMT*). In addition, the pzc determined using this method is independent of the charging mechanism assumed for the oxide surface, of the structure of the double layer, and of the surface-site density.

Surface charge and the estimation of surface acidity constants

The net surface charge, Q resulting from the ionization processes described by equations 1 and 2, may be defined as the difference in surface concentration between the positive and negative sites (see Davranche, 2003):

$$Q, \text{mol g}^{-1} = \{MOH^+_{2\,(surf)}\} - \{MO^-_{(surf)}\} \quad (9)$$

If one titrates the solid suspension with a monoprotic acid (e.g., HNO_3), at each point of the titration the following holds:

$$Q = (C_a - [H^+])/m \quad (10)$$

where m is the total amount of solid divided by the total volume of the suspension, and C_a is the concentration of the added acid.

By the same token, if the titration is done with a monobasic hydroxide (e.g., NaOH), at each point of the titration one has

$$Q = (-C_b + [OH^-])/m \quad (11)$$

where C_b is the concentration of the added base.

From equations 1 and 2 it is clear that the total concentration of surface sites, $\{S\}_{tot}$ is the sum of

the three types of possible sites,

$$\{S\}_{tot} = \{MOH\} + \{MO^-\} + \{MOH^+_2\} \quad (12)$$

$\{S\}_{tot}$ can be experimentally determined by measuring for example the amount of cations removed from solution by the target solid suspended in it (see Davranche, 2003).

In order to determine the values of the surface acidity constants (equations 5 and 6), one can discriminate between the following two cases:

(a) **pH values below the pzc**

Here the protonated sites predominate and $\{MO^-\}$ is negligible. Then,

$$\{S\}_{tot} = \{MOH\} + \{MOH^+_2\} \quad (13)$$

Because $\{MOH\}$ is neutral, the total surface charge is due to MOH^+_2 only:

$$Q = \{MOH^+_2\} = \{S\}_{tot} - \{MOH\} \quad (14)$$

Then, by substitution in equation 5 one can obtain K_{a1} in terms of measurable quantities:

$$K_{a1} = (\{MOH\}[H^+])/Q$$
$$= (\{S\}_{tot} - \{MOH^+_2\})[H^+]/Q$$
$$= (\{S\}_{tot} - Q)[H^+]/Q \quad (15)$$

(b) **pH values above the pzc**

Here the unprotonated sites predominate and $\{MOH^+_2\}$ is negligible. Then,

$$\{S\}_{tot} = \{MOH\} + \{MO^-\} \quad (16)$$

Because $\{MOH\}$ is neutral, the total surface charge is due to MO^- only:

$$Q = \{MO^-\} \quad (17)$$

Then, by substitution in equation 6 one can obtain K_{a2} in terms of measurable quantities:

$$K_{a2} = (\{MO^-\}[H^+])/\{MOH\}$$
$$= Q[H^+]/(\{S\}_{tot} - Q) \quad (18)$$

(note that the calculation of K_{a1} and of K_{a2} requires the **absolute value** of Q).

The values of K_{a1} and of K_{a2} vary with pH in an approximate linear fashion, and the following relationship has been inferred for the corresponding pK values:

$$pK_a = pK_{a(int)} + \beta Q \quad (19)$$

where $pK_{a(int)}$ represents the intrinsic value of this quantity (i.e., the pK_a at the pzc) and β is the slope.

From the values of pK_a the value of pzc can be calculated (see a related problem in the Postlaboratory questions and problems).

We present next a very simple experimental method to obtain the approximate pzc for several environmentally-relevant, simple metal oxides (TiO_2, SiO_2, Al_2O_3, ZnO, and MgO). These experiments have the objective of facilitating the comprehension of the phenomena involved in many surface processes involving simple oxides.

Experimental Procedure

Estimated time to complete the experiment: 4 h.

Materials	Reagents
1 pH meter	Al_2O_3
1 1-mL graduated pipet	TiO_2
1 2-mL graduated pipet	MgO
1 5-mL graduated pipet	ZnO
5 20-mL beakers	SiO_2
1 magnetic stirrer	0.1 M KNO_3
1 microburet	0.01 M KOH
1 universal stand	D. I. water
2 10-mL syringes	0.01 M HNO_3
1 propipet bulb	

Use the following procedure for each oxide. Commercial oxide reagents can be used. Aluminum oxide for column chromatography can also be used. Make sure to stir for at least 1 minute after every experimental step before taking and recording the pH measurements that involve such oxides. The longer the time allowed before a measurement, the closer the system will be to the true equilibrium. Depending on the accuracy desired, the experiments may be replicated as necessary.

(a) pzc determination of TiO_2, MgO, ZnO and Al_2O_3

Prepare two solutions by mixing 3.0 mL of 0.1 M KNO_3 and 6.0 mL of deionized (D.I.) H_2O in each of two 20-mL beakers. Measure the pH of either one. The first one will be called the *blank* and the other, the *sample*. Add to the *blank* 1.0 mL of 0.01 M KOH and measure the pH. Add to the *sample* 50 mg of the solid oxide and measure the resulting pH. Add 1.0 mL of 0.01 M KOH to the *sample* and measure the pH. Titrate the *blank* with HNO_3 0.01 M, and record the resulting curve. Do the same for the *sample*, and plot both curves in the same graph.

(b) pzc determination of SiO_2

The case of SiO_2 is different to those analyzed above due to its acidic nature, which increases the amount of OH^- required to deprotonate the initial sites. In addition, the ratio of the surface acidity of SiO_2 to its acidity in aqueous solution is much lower than that for many other solid oxides. This means that more OH^- are required than normal to neutralize its aqueous suspensions. To determine the pzc of SiO_2, prepare two solutions by mixing 3.0 mL of 0.1 M KNO_3 and 5.0 mL of D.I. H_2O in each of two 20-mL beakers. Measure the pH of either one. Add to the *blank* 2.0 mL of 0.01 M KOH and measure the pH. Add to the *sample* 50 mg of the solid oxide and measure the pH. Then, add 2.0 mL of 0.01 M KOH to the *sample* and measure the pH. Titrate the *blank* with HNO_3 0.01 M, and record the resulting curve. Do the same for the *sample*, and plot both curves in the same graph.

Name_____Section_____Date_____

Instructor_____Partner_____

PRELABORATORY REPORT SHEET—EXPERIMENT 5

Experiment Title _____

Objectives

Flow sheet of procedure

Waste containment and recycling procedure

PRELABORATORY QUESTIONS AND PROBLEMS

1. Look-up three applications of the concept of pzc and explain them in the light of the theory presented in the introduction.
2. In which cases is the *point of zero proton charge, pzpc* (or *zero proton condition, zpc*) different from the *point of zero charge*? Explain.
3. Predict the relative order of the pzc values of the metal oxides studied in this experiment on the basis of the ratio Z/R, where Z is the ionic charge of the cation and $R = 2r_0 + r_+$ (here, r_0 is the ionic radius of the oxide ion, and r_+ is the ionic radius of the cation). The greater this ratio, the more acidic the oxide, and the lower the pzc.

Additional Related Projects

• Determine the pzc of a unknown solid metal (hydr)oxide by mass titration. For this, add small amounts of the solid to an aqueous solution and measure the pH. The pzc is the point where an additional amount of solid does not produce further pH change. (See Preocanin, 1998).

- Determine the heat of immersion of an unknown solid metal (hydr)oxide and use a known correlation (see Healy, 1965) to estimate its pzc.
- Determine the pzc of an unknown solid metal (hydr)oxide by a differential potentiometric titration. To this end, titrate a suspension of the solid and plot $[H^+]$ vs. pH. At the pzc there is an inflexion point. If one takes the differential curve, there is an easily recognizable maximum at that point. (See Bourikas, 2005).

- Determine experimentally the total concentration of surface sites, $\{S_{tot}\}$ of an unknown solid metal (hydr)oxide by allowing complex cobalt hexaammine ions to exchange with the cations in the solid. Quantify the amount of complex remaining in solution after equilibration. (See Davranche, 2003).
- Determine the surface charge, Q of an unknown solid metal (hydr)oxide by titration as a function of pH. (See Davranche, 2003).
- Calculate the surface acidity constants, K_{a1} and K_{a2} from the titration data obtained in the previous project. (See Davranche, 2003).

Name_____Section_____Date_____

Instructor_____Partner_____

LABORATORY REPORT SHEET—EXPERIMENT 5

a) Plot the titration curves (pH vs Vol. HNO_3) for TiO_2, MgO, ZnO and Al_2O_3 and the corresponding blank curve.

b) Plot the titration curve for SiO_2 and the corresponding blank curve.

Volume of the supporting electrolyte   _____mL

Initial pH of the supporting electrolyte _____

c) Using your results from the points a) and b), fill-in the following table:

Oxide	Amount used, mg	Supporting electrolyte			Initial pH (after mixing)	Experim. pzc	Accepted pzc
		Conc.	Vol.	pH			

POSTLABORATORY PROBLEMS AND QUESTIONS

*1. The surface-ionization models (or surface-complexation models) account for the behavior of solid oxide suspensions, which usually behave as amphoteric substances. The most common are given in Table 1 (see Bourikas, 2005; Blesa, 1997; Davranche, 2003).

TABLE 1. Surface-ionization models (δ and ε are fractional charges).

Name	Surface ionization model	Equilibrium constants
One site / one pKa	$M - OH^{\delta+} \rightleftarrows M - O^{(1-\delta)-} + H^+$	K_a
One site / two pKa	$M - OH_2^+ \rightleftarrows M - OH + H^+$	K_{a1}
	$M - OH \rightleftarrows M - O^- + H^+$	K_{a2}
Two sites / two pKa	$M - OH_2{}^{\delta+} \rightleftarrows M - OH^{(1-\delta)-} + H^+$	K_{a1}
	$M - OH^{(1-\varepsilon)+} \rightleftarrows M - O^{\varepsilon-} + H^+$	K_{a2}
Multisite	$M_{xi} - OH_{yi}{}^{zi} \rightleftarrows M_{xi} - OH_{yi-1}^{zi-1} + H^+$	K_i

As an example of calculation, let us analyze the first model. Here,

$$K_a = \frac{[M - O^{(1-\delta)-}][H^+]}{[M - OH^{\delta+}]}$$

The point where the total sum of positive and negative changes is zero (i.e., the pzc) is achieved in this model when $[M - O^{(1-\delta)-}] = [M - OH^{\delta+}]$, since we assume that the protons diffuse away. Then, $K_a = [H^+]$ and $pK_a = pH = pzc$.

Assuming again that the protons dissolve away and that electroneutrality must be achieved at the solid sphere, answer the following for the second model (i.e., one site/two pK_a):

(a) How does $[M - OH_2^+]$ compare with $[M - O^-]$?
(b) Write the pH at which pzc is achieved, in terms of pK_{a1} and pK_{a2}.

*2. In a real test, 10 g of a river sediment were added to a 0.1 M solution of a supporting electrolyte (e.g., $NaNO_3$) to a final volume of 100 mL. The resulting suspension was titrated at first with HNO_3 and then with NaOH. By ion exchange, the total surface sites were calculated as $\{S\}_{tot} = 0.00013$ mol/g. Selected data points are given in Table 2. (See Davranche, 2003).

With this information,

(a) Calculate the value of Q at each point
(b) Plot Q vs pH and estimate the value of pzc
(c) Calculate the value of K_{a1} at each point below the pzc
(d) Calculate the value of K_{a2} at each point above the pzc
(e) Calculate the *intrinsic values* of pK_{a1} and pK_{a2}

TABLE 2. Experimental titration values from a river sediment

V, mL HNO₃	pH, meas.
0.1	5.33
0.2	4.86
0.5	4.47
0.8	4.02
1	3.72

V, mL NaOH	pH, meas.
0.1	6.92
0.2	7.07
0.5	7.56
0.8	7.93
1	8.13

(f) With the values obtained from e), calculate the pzc and compare it with that obtained in (b).

*3. The surface acidity of a metal oxide may be considered as *Lewis acidity*, which is a result of the electron-accepting character of the oxide surface. The more acidic the surface, the fewer the acidic sites to neutralize and the lower the pzc of the oxide. Since the ionization of a metal M to form the metal ion M^{z+} in the oxide can be written as

$$M + I = M^{z+} + Ze^-$$

where I is the (total) ionization energy of the element (in kJ/mole), then it follows that the larger the I, the higher the acidity of the resulting cation since it will have a greater tendency to return to its original state by attracting electrons.

Test for this correlation with your own results by plotting your *pzc* vs. *I*. (Look up the different ionization energies in a handbook or elsewhere). For example, the following relationship was found in the literature (see Carre, 1992):

$$pzc = 12.2 - 0.0009\,I$$

Student Comments and Suggestions

Literature References

Balderas-Hernandez, P.; Ibanez, J. G.; Godinez-Ramirez, J.J.; Almada-Calvo, F. "Microscale Environmental Chemistry: Part 7. Estimation of the Point of Zero Charge (pzc) for Simple Metal Oxides by a Simplified Potentiometric Mass Titration Method," *Chem. Educ.* **2006**, *11*, 267–270.

Barthés-Labrousse, M.-G. "Acid-Base Characterization of Flat Oxide-Covered Metal Surfaces," *Vacuum* **2002**, *67*, 385–392.

---

*Answer in this book's webpage at www.springer.com

Blesa, M. A.; Magaz, G.; Salfity, J. A.; Weisz, A. D. "Structure and Reactivity of Colloidal Metal Oxide Particles Immersed in Water," *Solid State Ionica* **1997**, *101–103*, 1235–1241.

Bourikas, K.; Kordulis, C.; Lycourghiotis, A. "Differential Potentiometric Titration: Development of a Methodology for Determining the Point of Zero Charge of Metal (Hydr)oxides by One Titration Curve," *Environ. Sci. Technol.* **2005**, *39*, 4100–4108.

Bourikas, K.; Vakros, J.; Kordulis, C.; Lycourghiotis, A. "Potentiometric Mass Titrations: Experimental and Theoretical Establishment of a New Technique for Determining the Point of Zero Charge (PZC) of Metal (Hydr)Oxides," *J. Phys. Chem. B*, **2003**, *107*, 9441–9451.

Carre, A.; Roger, F.; Varinot, C. "Study of Acid/Base Properties of Oxide, Oxide Glass, and Glass-Ceramic Surfaces," *J. Coll. Interf. Sci.* **1992**, *154*, 174–183.

Davranche, M.; Lacour, S.; Bordas, F.; Bollinger, J.-C. "An Easy Determination of the Surface Chemical Properties of Simple and Natural Solids," *J. Chem. Educ.* **2003**, *80*, 76–78.

Hamieh, T. "Etude des Proprietes Acido-Basiques et Energie Interfaciale des Oxides et Hydroxides Metalliques," *C. R. Acad. Sci. Paris (Chim. Surf. Catal.*, in French) **1997**, *325 (Serie IIb)*, 353–362.

Healy, T. W.; Fuerstenau, D. W. "The Oxide-Water Interface–Interrelation of the Zero Point of Charge and the Heat of Immersion," *J. Coll. Sci.* **1965**, *20*, 376–386.

Kraepiel, A. M. L.; Keller, K.; Morel, F. M. M. "On the Acid-Base Chemistry of Permanently Charged Minerals," *Environ. Sci. Technol.* **1998**, *32*, 2829–2838.

Mullet, M.; Fiebet, P.; Szymczyk, A.; Foissy, A.; Reggiani, J.-C.; Pagetti, J. "A Simple and Accurate Determination of the Point of Zero Charge of Ceramic Membranes," *Desal.* **1999**, *121*, 41–48.

Reymond, J. P.; Kolenda, F. "Estimation of the Point of Zero Charge of Simple and Mixed Oxides by Mass Titration," *Powder Technol.* **1999**, *103*, 30–36.

Preocanin, T.; Kallay, N. "Application of Mass Titration to Determination of Surface Charge of Metal Oxides," *Croat. Chem. Acta.* **1998**, *71*, 1117–1125.

Sahai, N. "Is Silica Really an Anomalous Oxide? Surface Acidity and Aqueous Hydrolysis Revisited," *Environ. Sci. Technol.* **2002**, *36*, 445–452.

Vakros, J.; Kordulis, C.; Lycourghiotis, A. "Potentiometric Mass Titrations: A Quick Scan for Determining the Point of Zero Charge," *Chem. Comm.* **2002**, 1980–1981.

Yoon, R. H.; Salman, T.; Donnay, G. "Predicting Points of Zero Charge of Oxides and Hydroxides," *J. Coll. Interf. Sci.* **1979**, *70*, 483–493.

Experiment 6
Experimental Transitions in E vs pH (or Pourbaix) Diagrams

Reference Chapters: 2, 6, 10

Objectives

After performing this experiment, the student shall be able to:

- Verify experimentally many of the transitions in metal Pourbaix diagrams.
- Understand the role of potential in metal speciation.
- Understand the role of pH in metal speciation.

Introduction

Different metal species may predominate in a given aqueous system depending on its pH and redox potential. To summarize such predominances, Pourbaix devised a graphical representation that facilitates the prediction of the most stable species under a given set of E and pH values, as discussed in Section 2.3. Those reactions involving only electron exchange are E-dependent and when plotted they give horizontal lines, whereas those involving H^+ or OH^- (but not e^-) are pH-dependent and give vertical lines. Equilibrium calculations for reactions involving both (e^- and H^+/OH^- exchange) yield inclined lines.

In the present experiment, several chemical and electrochemical transitions—some including the metal itself—can be predicted and verified using simple reagents and in uncomplicated conditions (see Ibanez et al., 2005). Copper was selected because of the different colors of its compounds, its relatively low toxicity, wide availability in fairly pure form, and low cost. Ten different transitions

in the E-pH diagram can be readily observed, as described below. It is important to respond in advance the second question in the prelaboratory assignment, see below. (Note that the equations and the diagram given there assume 1 M solutions, whereas the experiment is suggested for more dilute conditions. Also note that due to the qualitative nature of the experiment, the amounts and parameters given below, in the experimental procedure, are to be taken only as a guide and thus need not be reproduced exactly).

Experimental Procedure

Estimated time to complete the experiment: 2.5 h.

Materials	Reagents
1 pH meter	0.1 M $CuSO_4$
1 well-plate (6)	0.1 M Na_2SO_4
5 Beral pipets	1 M NaOH
1 1-mL graduated pipet	1 M H_2SO_4
1 2-mL graduated pipet	8 M NaOH
1 U-tube	1 M HCl
1 5-mL graduated pipet	50 % HNO_3
1 magnetic stirrer	1 Zn pellet
1 spectrophotometer	D. I. water
1 5-mL beaker	
1 hot plate	
1 - 9 V battery, or a DC converter	
2 alligator clips	
electrodes (graphite rods, stainless steel and Cu wire)	
filter paper	
1 propipet bulb	

(a) Chemical transitions

For these chemical transitions, use a plastic plate with at least six wells (with a minimum capacity of about 3 mL each). If desired, a thin-stem pH probe connected to a pH meter may be used to monitor the pH in the wells (although this reduces somewhat the volume available for the tests; 5- or 10-mL beakers may be used instead of the well-plate). If even smaller volumes are desired, use a porcelain spot test plate. For color comparison purposes, place 2 mL of 0.1 M $CuSO_4$ in the first well.

1. Cu^{2+} to $Cu(OH)_2$

Place 2 mL of 0.1 M $CuSO_4$ in the second well and add dil. NaOH drop wise. When the pH reaches approximately 4.5, a sky-blue precipitate of $Cu(OH)_2$ forms. At this point, stop adding NaOH. Save the resulting mixture of solution and precipitate.

2. $Cu(OH)_2$ to CuO_2^{2-}

Take approximately 1 mL of the reaction mixture produced in the previous step and place it in the third well. Add a few drops of a highly concentrated (e.g., 8 M) NaOH solution. Stir. **Caution: NaOH dissolution is an exothermic process, and concentrated NaOH is very caustic. Eye protection must be worn.** A purplish deep-blue color indicates the production of cuprite ion, CuO_2^{2-}. Keep this solution for later. If desired, the absorption spectrum of cuprite ($\lambda_{max} = 620$ nm) can be compared to that of Cu^{2+} ($\lambda_{max} = 811$ nm).

3. CuO_2^{2-} to $Cu(OH)_2$

Remove half of the contents of the third well and add dilute H_2SO_4 drop wise (with stirring) to the mixture remaining in it to a sky-blue color, which indicates the formation of $Cu(OH)_2$ precipitate.

4. $Cu(OH)_2$ to Cu^{2+}

Upon further addition (with stirring) of H_2SO_4 to the mixture remaining in the third well, the $Cu(OH)_2$ precipitate starts dissolving and a normal blue color appears, indicating that copper is back to its initial form (i.e., Cu^{2+}).

5. $Cu(OH)_2$ to CuO

Take approximately 1 mL of the mixture from the second well, put it in a 5-mL beaker, and place it on a hot plate. Make sure that the pH

of the reaction mixture is above 9. Then, heat so as to boil-off the excess water and to dehydrate the hydroxide. A black precipitate of CuO now forms as a result of the hydroxide dehydration. **Caution: Some splashing may occur. Wear safety goggles.** If desired, the CuO thus produced can be treated with concentrated acid (e.g., 50% HNO_3) as to yield blue Cu^{2+}. If the CuO is treated with a strong base (e.g., a few drops of conc. NaOH and a few drops of H_2O) and heated, the mixture yields the deep-blue CuO_2^{2-} ion. (Some time must be allowed in order to notice the blue color). **Caution: NaOH dissolution is an exothermic process. Concentrated nitric acid emits toxic fumes, handle under a fume hood. Wear gloves and safety goggles.**

(b) Electrochemical transitions
To observe the electrochemical transitions, use the remaining wells on the well-plate. Dip into each well the required electrodes (e.g., 5 cm of stainless steel wire, SS, Cu wire, or graphite rod) as described below.

6. Cu^{2+} to Cu^0 (electrolytic)

Place 2 mL of 0.1 M $CuSO_4$ in the fourth well, and adjust the pH to 1-2 with dil. sulfuric acid. Place two SS wires (cathode and anode, respectively) in the solution. (Make sure that they do not touch each other). Connect them to a 9 V battery or to a DC converter set at approximately 9 V (this is the *power source*). Within a few minutes, the cathode will be covered with a reddish deposit of Cu^0. To test for its presence, remove the cathode and dip the reddish portion in a small test tube (e.g., 5 cm long, 0.5 mm diameter) more than half-filled with 50% HNO_3 for a few seconds. Brownish fumes of NO_2 are indicative of the presence of Cu^0. **Caution: Do this step under a fume hood. Use safety goggles and gloves to handle the 50% HNO_3. These precautions are imperative because NO_2 fumes are toxic, and HNO_3 is highly corrosive to human tissue.**

7. Cu^{2+} to Cu^0 (cementation)
Place 3 mL of 0.1 M $CuSO_4$ in the fifth well of the well-plate. Add Zn metal (either a few mg of dust or a pea-sized pellet). After some minutes, observe the Cu^0 deposit on the Zn surface as

well as the discoloration of the blue solution. Compare the color of the solution with that of the solution in the first well.

8. Cu^{2+} to Cu_2O

Place 3 mL of 0.1 M CuSO$_4$ in the sixth well and *carefully* adjust the pH to approximately 3.5 with dil. sulfuric acid or hydrochloric acid (and NaOH if needed). Connect two SS wires to each pole of the power source. Stir. Before long, an opaque reddish deposit forms on the cathode surface. To make sure that the deposit is not Cu^0, remove the cathode and dip the reddish portion in a few milliliters of 50% HNO$_3$ as in step 6. Brownish fumes of NO$_2$ are not emitted here since it is not Cu^0 that was formed, but Cu_2O. **Caution: same as in the aforementioned step.** An alternative test can be performed in the same way, but using dil. HCl (for example, 0.05 M) instead of 50% HNO$_3$. Here, the Cu_2O deposit dissolves, whereas Cu^0 metal would not.

9. Cu^0 to Cu^{2+}

In a small beaker, place 2 mL of distilled or deionized water, and add 10 drops of 0.1 M Na$_2$SO$_4$ as electrolyte. Adjust the pH to 1 with dil. H$_2$SO$_4$. Place the resulting solution in a small U-tube (e.g., 6 mm inside diameter, 5 cm tall). Insert the electrodes (a graphite rod or a SS wire as the cathode, and a copper wire of any gauge as the anode) on each side arm of the U-tube. Connect the electrodes to the power source (the Cu wire to the positive). See Figure 1 (the marker seen there is simply used to hold the connectors with Scotch tape). After 10–15 minutes, a faint normal blue color appears, indicating the presence of Cu^{2+} ions coming from the oxidation of the Cu^0 electrode. If desired, these ions can be made more visible by adding a dilute NH$_4$OH solution drop wise.

Note: Color production is faster with a larger diameter Cu wire; alternatively, make a "paper ball" from a piece of filter paper, wet it with electrolyte, and insert it as a separator at the bottom of the U-tube. Yet a third possibility consists

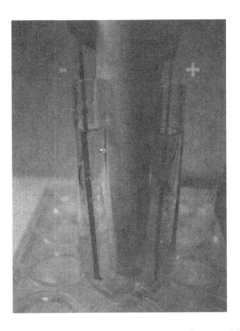

FIGURE 1. Alternative experimental cell for transition number 9. (Reprinted from the Chemical Educator, Vol. 10, No. 5, 2006, pp. 348–356, with permission).

of pouring the solution with the Cu^{2+} newly produced into the preparation beaker and add NH$_4$OH as needed to form the highly-colored $[Cu(NH_3)_2]^{2+}$ complex.

10. Cu^0 to CuO_2^{2-}

Place 2 mL of 8 M NaOH in a 5- or 10-mL beaker. Insert a copper wire as the anode and either a graphite rod or a SS wire as the cathode. (To observe the Cu^0 to CuO_2^{2-} transition in a relatively short period of time, it is best to connect two or three 9 V batteries in parallel as to have the same voltage, but a higher current). Then, connect the electrodes to the power source (the copper wire to the positive). In a few minutes a deep-blue color forms in the solution near the anode surface. **Caution: NaOH is very corrosive to skin tissue and to the eyes. Use safety goggles and gloves.**

All the solutions generated in this experiment must be placed in a container labeled for this purpose.

Name_____Section_____Date_____

Instructor_____Partner_____

PRELABORATORY REPORT SHEET—EXPERIMENT 6

Experiment title_____

Objectives

Flow sheet of procedure

Waste containment and recycling procedure

PRELABORATORY QUESTIONS AND PROBLEMS

*1. Knowing that Zn(II) hydroxide is amphoteric and that it forms divalent ions in acidic and in basic solutions,

a) Replace the alphabet letters in the boxes in Figure 2 with the chemical formulas of the missing species.

b) Write the (balanced) equations that describe each one of the equilibrium lines in this diagram (except the dotted ones, that represent the water equilibria; these two lines do not participate directly in the Zn equilibria).

c) Justify chemically and/or algebraically the type of slope in each Zn equilibrium line.

*2. Calculate and draw the Pourbaix diagram for Cu in the E range of −2 to +2 V vs NHE (normal hydrogen electrode), and pH range 0–20

at 1 M concentration of dissolved species, 25°C, and 1 atm. You may draw a simplified diagram, considering only the following Cu species: (a) dissolved species (Cu^+, Cu^{2+}, CuO_2^{2-}), and (b) solid species [Cu, $Cu_2O_{(s)}$, $Cu(OH)_{2(s)}$]. The equations and the associated thermodynamic data necessary to calculate this diagram are given below. The calculations assume that the solutions do not contain dissolved oxygen. For the sake of simplicity, the physical state of the aqueous species is not specified.

Note: If you have not studied the construction of these diagrams, you may obtain one from the literature and insert it here.

$$Cu_2O_{(s)} + H_2O_{(1)} \rightleftarrows 2Cu^+ + 2OH^-$$
$$pK = 29.4 \tag{1}$$
$$Cu(OH)_{2(s)} \rightleftarrows Cu^{2+} + 2OH^-$$
$$pK_{sp} = 19.6 \tag{2}$$

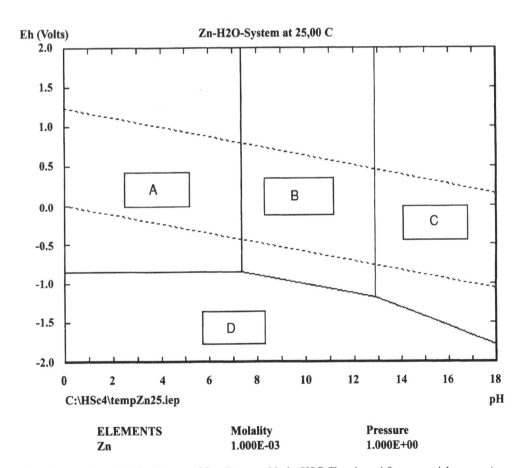

FIGURE 2. A simplified Pourbaix diagram of Zn. (Drawn with the HSC Chemistry 4.0 commercial program).

$$Cu(OH)_{2(s)} \rightleftharpoons CuO_2^{2-} + 2H^+$$

$$pK_{sp} = 32.4 \qquad (3)$$

$$Cu^+ + e^- \rightleftharpoons Cu_{(s)}$$

$$E^0 = 0.52 \text{ V} \qquad (4)$$

$$Cu_2O_{(s)} + 2H^+ + 2e^- \rightleftharpoons 2Cu_{(s)} + H_2O_{(1)}$$

$$E^0 = 0.46 \text{ V} \qquad (5)$$

$$Cu^{2+} + e^- \rightleftharpoons Cu^+$$

$$E^0 = 0.16 \text{ V} \qquad (6)$$

$$2Cu^{2+} + H_2O_{(1)} + 2e^- \rightleftharpoons Cu_2O_{(s)} + 2H^+$$

$$E^0 = 0.22 \text{ V} \qquad (7)$$

$$2Cu(OH)_{2(s)} + 2H^+ + 2e^- \rightleftharpoons Cu_2O_{(s)} + 3H_2O_{(1)}$$

$$E^0 = 0.73 \text{ V} \qquad (8)$$

$$2CuO_2^{2-} + 6H^+ + 2e^- \rightleftharpoons Cu_2O_{(s)} + 3H_2O_{(1)}$$

$$E^0 = 2.67 \text{ V} \qquad (9)$$

$$CuO_2^{2-} + 4H^+ + 2e^- \rightleftharpoons Cu_{(s)} + 2H_2O_{(1)}$$

$$E^0 = 1.56 \text{ V} \qquad (10)$$

Note that $Cu_{(aq)}^+$ does not appear in the diagram because its disproportionation reaction into $Cu_{(aq)}^{2+}$ and $Cu_{(s)}$ is spontaneous, as can be calculated from equations 4 and 6.

*3. a) Balance the reactions for the oxidation and reduction of water (both written as reductions). Note that in the reduction of water, the species that becomes reduced is the hydrogen and not the oxygen; thus, the reduction reaction is often written as the reduction of protons:

$$O_{2(g)} + H^+ + e^- \rightleftharpoons H_2O_{(1)}$$

$$H^+ + e^- \rightleftharpoons H_{2(g)}$$

b) Write the two algebraic equations that relate the standard potential of each reaction and pH, as derived from the Nernst equation. Make the assumptions and simplifications that you may deem reasonable. Use $T = 25°C$, 1 atm, concentration of all dissolved species $= 1$ M. You may use $(RT/F) \ln x = 0.06 \log x$, $RT/F = 0.0257$, $E^0{}_{H^+/H_2} = 0.00$ V and $E^0{}_{O_2/H_2O} = +1.23$ V.

c) With these data, draw the E-pH (Pourbaix) for water.

Additional Related Projects

• Observe transitions among different regions involving the production or disappearance of V or Co species, as the pH and/or E vary in an aqueous solution. (See Powell, 1987).

• Predict the change in standard potentials of some metal ion couples upon complexation with selected ligands. For example, predict and then test the oxidizing or reducing power of the $Fe^{3+/2+}$ couple with a solution of I^- and a solution of I^-/I_2. Now add some EDTA to these mixtures. Analyze your observations in the light of the new standard potential of the metal ion couple upon complexation (see Section 2.4 and problem 2 above). See Napoli, 1997 and Ibanez, 1988.

• Calculate and draw the Pourbaix diagram for iron. Compare it to those drawn by Pourbaix, 1974 and Barnum, 1982.

---

* Answer in the book's webpage at www.springer.com

Name_____Section_____Date_____

Instructor_____Partner_____

LABORATORY REPORT SHEET—EXPERIMENT 6

Concentration of the Cu(II) solution utilized _____M

a) *Chemical transitions*

1. Cu^{2+} to $Cu(OH)_2$

Color before the addition of NaOH _____

Color after the addition of NaOH _____

Final pH _____

2. $Cu(OH)_2$ to CuO_2^{2-}

Color before the addition of NaOH _____

Color after the addition of NaOH _____

Final pH _____

Peak at the absorption maximum of Cu(II) _____nm

Peak at the absorption maximum of $CuO_2{}^{2-}$ _____nm

3. $CuO_2{}^{2-}$ to $Cu(OH)_2$

Color before the addition of H_2SO_4 _____

Color after the addition of H_2SO_4 _____

Final pH _____

4. $Cu(OH)_2$ to Cu^{2+}

Color before the addition of H_2SO_4 _____

Color after the addition of H_2SO_4 _____

Final pH _____

5. $Cu(OH)_2$ to CuO

Color before heating _____

Color after heating _____

b) Electrochemical transitions

6. Cu^{2+} to Cu^0 (electrolytic)

Anode material   _____

Cathode material   _____

Voltage applied   _____V

Observations upon reaction with HNO_3 _____

7. Cu^{2+} to Cu^0 (cementation)

Observations   _____

8. Cu^{2+} to Cu_2O

Anode material   _____

Cathode material   _____

Voltage applied   _____V

Observations upon reaction with HNO_3, HCl   _____

9. Cu^0 to Cu^{2+}

Anode material   _____

Cathode material   _____

Voltage applied   _____

Observations   _____

10. Cu^0 to $CuO_2{}^{2-}$

Anode material   _____

Cathode material   _____

Voltage applied   _____V

Color before oxidation   _____

Color after oxidation   _____

POSTLABORATORY PROBLEMS AND QUESTIONS

*1. In the diagram below (Figure 3), the location of several natural environments is given. Assign each one to the appropriate letter: organic-rich saline waters, water-logged soils, aerated saline waters, mine waters. (Note: The dotted lines represent water equilibria).

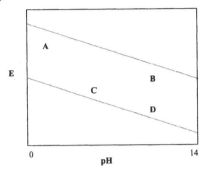

FIGURE 3. The location of several natural environments in a Pourbaix diagram.

*2. As discussed in Section 2.4, the standard potential of metal ions can be significantly altered upon complexation. For the equilibrium between M^{m+} and $M^{(m-n)+}$, represented by

$$M^{m+} + ne^- = M^{(m-n)+}$$

with a standard potential of E^0_{aq} (see Napoli, 1997 and Ibanez, 1988), the line that governs such equilibrium is independent of pH and a portion of it is given in Figure 4.

If both ions now react with the non-charged ligand L, a new equilibrium is established:

$$ML^{m+} + ne^- = ML^{(m-n)+}$$

Assuming that ML^{m+} is more stable than $ML^{(m-n)+}$, and that L does not participate in acid-base nor hydrolysis equilibria, which of the options in Figure 5 represents this new equilibrium?

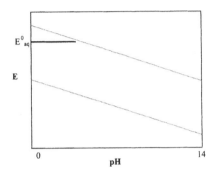

FIGURE 4. The equilibrium line between M^{m+} and $M^{(m-n)+}$ represented by $M^{m+} + ne^- = M^{(m-n)+}$ with a standard potential $= E^0_{aq}$.

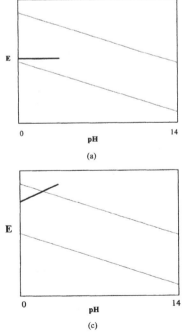

(a)

(c)

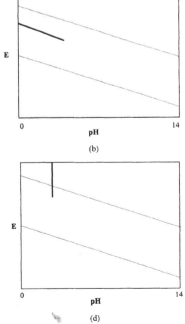

(b)

(d)

FIGURE 5. E-pH response of a metal ion upon complexation.

*3. Sketch the predominance-zone diagram for the $Cu^{2+}/Cu(OH)_2/CuO_2^{2-}$ system using the conditions and equilibrium data from Problem 2 in the Prelaboratory questions and problems.

Student Comments and Suggestions

Literature References

Barnum, D. W. "Potential-pH Diagrams," *J. Chem. Educ.* **1982**, *59*, 809–812.

Campbell, J. A.; Whiteker, R. A. "A Periodic Table Based on Potential-pH Diagrams," *J. Chem. Educ.* **1969**, *46*, 90–92.

HSC Chemistry 4.0. Outokumpu Research OY, PO Box 60, FIN-28101, Pori, Finland.

Ibanez, J. G. "Redox Chemistry and the Aquatic Environment: Examples and Microscale Experiments," *Chem. Educ. Int. (IUPAC)* **2005**, *6*, 1–7.

Ibanez, J. G.; Garcia, K.; Balderas-Hernandez, P. "Microscale Environmental Chemistry, Part 4. Experimental Transitions in a Potential vs. pH or Pourbaix Diagram," *Chem. Educator* **2005**, *10*, 348–351.

Ibanez, J. G.; González, I.; Cárdenas, M. A. "The Effect of Complex Formation upon the Redox Potentials of Metallic Ions," *J. Chem. Educ.* **1988**, *65*, 173–175.

King, D. W. "A General Approach for Calculating Speciation and Poising Capacity of Redox Systems with Multiple Oxidation States: Application to Redox Titrations and the Generation of pε-pH Diagrams," *J. Chem. Educ.* **2002**, *79*, 1135–1140.

Little, J. A. University of Cambridge. http://www.msm.cam.ac.uk/jal/C8-4/C8-4-1.htm.

Napoli, A.; Pogliani, L. "Potential-pH diagrams," *Educ. Chem.* March **1997**, 51–52.

Pourbaix, M. *Atlas of Electrochemical Equilibria in Aqueous Solutions,* 2nd English ed.; National Association of Corrosion Engineers: Houston and CEBELCOR, Brussels. 1974.

Powell, D.; Cortez, J.; Mellon, E. K. "A Laboratory Exercise Introducing Students to the Pourbaix Diagram for Cobalt," *J. Chem. Educ.* **1987**, *64*, 165–167.

Singh, M. M.; Szafran, Z.; Pike, R. M. "Microscale and Selected Macroscale Experiments for General and Advanced General Chemistry: An Innovative Approach," Wiley: NY, 1995. Chapter 7.

Tykodi, R. J. "In Praise of Copper," *J. Chem. Educ.* **1991**, *68*, 106–109.

Williams, B. G.; Patrick, W. H. Jr. "A Computer Method for the Construction of E*h*-pH Diagrams," *J. Chem. Educ.* **1977**, *54*, 107.

---

* Answer in this book's webpage at www.springer.com

Experiment 7
Air Oxidation of Metal Ions

Reference Chapters: 2, 5, 6, 8

Objectives

After performing this experiment, the student shall be able to:

- Explain how the solution pH affects the extent of oxidation of iron(II) to iron(III) by dioxygen.
- Interpret the changes observed during the course of the reactions.
- Apply a potentiometric titration as an analytical technique to study a heterogeneous reaction.

Introduction

The oxidation state of a chemical species is often responsible for its environmental behavior. For instance, it is well known that As(III) compounds tend to be very toxic, while the opposite is true for the more oxidized As(V) compounds. On the other hand, Cr(VI) ions (in the form of chromate or dichromate) are toxic and highly mobile in nature, whereas Cr(III) ions are much less toxic and easily immobilized as the corresponding hydroxide.

Natural or anthropogenic processes can often change an oxidation state. For example, the redox state of iron depends on the presence of dioxygen, light, and biocatalysts. Iron compounds are key components of the hydrosphere for several reasons: (a) they are present in virtually every water body and groundwater reservoir in the world, as well as in atmospheric water; (b) they play a key role in several natural chemical processes, such as their reaction with H_2O_2 to produce the highly oxidizing hydroxyl radicals; (c) they can be critical in certain biological processes (for example, in the phytoplankton growth in eutrophication phenomena and in microbial cycles).

The thermodynamic stability of Fe(II) under different conditions of pH and potential can be predicted from Latimer, Frost, or Pourbaix diagrams (see Chapter 2). Fe(II) is stable in anoxic water (i.e., water without oxidizing conditions), but the presence of oxygen drives the equilibrium toward Fe(III) and reactive oxygen species. Such Fe(III) species have low solubility, and the hydrous ferric oxide normally formed is capable of readily adsorbing other trace metals present, thus, reducing their concentration. In fact, biological oxidation of groundwater is purposefully promoted in order to form insoluble Fe(III) species that can be easily removed by mechanical means. On the other hand, Fe(II) is much more soluble and mobile than Fe(III). This is a key issue because iron bioavailability strongly depends on its solubility. Light-induced reduction of Fe(III) contributes to the formation of Fe(II) in different types of waters. Also, Fe(II) is the predominant form of dissolved iron in fog water at low pH, and its oxidation under such conditions (and in the absence of catalysts) can have half-lives in the order of two months.

The mechanism for the oxidation of Fe(II) by molecular oxygen is known as the Haber–Weiss mechanism and can be written as follows (see

Emmenegger, 1998):

$$Fe(II) + O_2 \xrightarrow{k_1} Fe(III) + O_2^- \qquad (1)$$

$$Fe(II) + O_2^- + 2H^+ \xrightarrow{k_2} Fe(III) + H_2O_2 \qquad (2)$$

$$Fe(II) + H_2O_2 \xrightarrow{k_3} Fe(III) + {}^\bullet OH + OH^- \qquad (3)$$

$$Fe(II) + {}^\bullet OH \xrightarrow{k_4} Fe(III) + OH^- \qquad (4)$$

Some of these reactions are kinetically limited, and so equilibria between the iron(II) and iron(III) oxidation states in real situations depend upon their interconversion rate. From the mechanism described by equations 1 through 4, the rate expression has been shown to be

$$\frac{-d[Fe(II)]}{dt} = 2k_1[O_2][Fe(II)] + 2k_3[H_2O_2][Fe(II)] \qquad (5)$$

(see To, 1999).

If $[O_2]$ and $[H_2O_2]$ are assumed to be constant, then

$$\frac{-d[Fe(II)]}{dt} = k_{app}[Fe(II)] \qquad (6)$$

where k_{app} is the apparent first-order rate constant that includes the O_2 and H_2O_2 concentrations.

Several other factors also affect the rate and the extent of Fe(II) oxidation. For instance, in the presence of bicarbonate ions (from the dissolution of dissolved atmospheric CO_2) this rate has been shown to be:

$$\frac{-d[Fe(II)]}{dt} = k[O_2][Fe(II)][OH^-]^2 \qquad (7)$$

(see To, 1999).

Speciation of Fe(II) is a key parameter influencing such a rate. The presence of some inorganic and organic ligands is also known to substantially affect this oxidation rate, as well as bacteria and the presence of solid Fe(III) particles that promote autocatalytic Fe(II) oxidation.

The importance of the Fe(II)/Fe(III) equilibrium is clear from all these considerations (see King, 1998). In other experiments in this book we address the complexation of NO with [Fe(II)EDTA], the photodecomposition of [Fe(III)EDTA], and the reaction of Fe(II) with H_2O_2 (Fenton's reaction). In the present experiment we explore the influence of pH upon the oxidation of Fe(II) by bubbling air through Fe(II) solutions at different pH values. In Part 1, a qualitative experiment aimed at showing the influence of pH upon the production of an Fe(III)

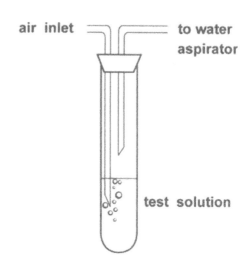

FIGURE 1. Experimental setup for the qualitative observation of the oxidation of Fe^{2+} ions by air.

precipitate is described. In Part 2, a more quantitative experiment will be performed by bubbling air simultaneously through a series of Fe(II) solutions at different pH values; the Fe(II) remaining in each solution at the end of the experiment will be analyzed by potentiometric titration.

Experimental Procedure

Estimated time to complete the experiment: 1.5 h

Materials	Reagents
1 pH meter, with ORP	0.05 M $FeSO_4 \cdot 7\,H_2O$
1 aquarium pump	1 M KSCN
10 cm Tygon tube	2 M HCl
1 1-mL graduated pipet	0.01 M $K_4[Fe(CN)_6]$
1 2-mL graduated pipet	2 M NH_4OH
10 test tubes, 10 cm long	concentrated HCl
5 Beral pipets	D. I. water
1 magnetic stirrer	5 M NaOH
4 rubber stoppers	0.1 M H_2SO_4
1 5-mL beaker	0.01 M H_2SO_4
Filtration membranes, 0.45 μ	3 M H_2SO_4
and supports	0.05 M $KMnO_4$
microburet	
1 universal stand	
2 10-mL syringes	
1 2-mL serologic pipet	
1 propipet bulb	

Caution: $FeSO_4$ is harmful. Do not touch it with your bare hands nor breathe its dust. Also, any cyanide-containing solution [e.g., hexacyanoferrate

(III), hexacyanoferrate (II)] could potentially release highly toxic cyanide vapors in the presence of a strong acid. Handle under a fume hood.

Part 1. Qualitative demonstration.

Prepare 4 to 5 mL of a 0.05 M $FeSO_4 \cdot 7H_2O$ solution. In a small test tube, take a 0.5 mL aliquot and add two drops of 2 M HCl. Then test for the presence of Fe(III) ions by adding one or two drops of a 1 M KSCN solution. A faint brown color is indicative of the absence of Fe(III) ions. Take another 0.5 mL aliquot, add two drops of 2 M HCl and one or two drops of dilute $K_4[Fe(CN)_6]$ (i.e., 0.1 M). A white-blue precipitate (known as Everitt's blue) is also indicative of the absence of Fe(III) ions and of the presence of Fe(II) ions.

The remaining $FeSO_4$ solution (approximately 3 mL) is then divided into two equal parts and placed in two 10-cm test tubes (tubes I and II). Add a 2 M NH_4OH solution drop wise to tube I until a greenish Fe(II) hydroxide precipitate forms (approximately at pH 5). Then, bubble air through both solutions (for example, with the aid of a water aspirator or an aquarium pump) for 15 min or until the precipitate in tube I changes color to orange-brown. See Figure 1.

Centrifuge the mixture (tube I) for approximately 2 min, and discard the supernatant liquid. Wash the precipitate by adding some 3–4 mL of water, and centrifuge again. Discard the supernatant liquid, and add to the precipitate 2 mL of concentrated HCl. Heat in a water bath until the precipitate dissolves, and allow the tube to cool down to room temperature. Take 1 mL of this solution (it should be yellow-orange), and test for the presence of Fe(III) ions as described above. A red color with the KSCN solution and an intense blue precipitate with $K_4[Fe(CN)_6]$ are indicative of the presence of Fe(III) ions.

Perform these same tests with the solution obtained in test tube II. A faint red color with KSCN indicates that there are some Fe(III) ions, albeit in a concentration much smaller than that in test tube I. A similar result is obtained with the $K_4[Fe(CN)_6]$ test where a small amount of blue precipitate indicates the presence of only a small concentration of Fe(III).

Part 2. Quantitative determination of the oxidation of Fe(II) by molecular oxygen and its precipitation as Fe(III).

Prepare several test tubes containing 3-mL aliquots of 0.05 M $FeSO_4 \cdot 7H_2O$ at different pH values between approximately 2 and 12. The number of test tubes may be assigned by the instructor or decided by the students. Adjust the pH in each aliquot with the smallest possible volume of 0.01 to 5 M NaOH or 0.1 or 0.01 M H_2SO_4, and then dilute to a final volume of 4 mL. (Note: in the solutions with pH 7 to 8, a precipitate starts forming). Obtain a fish-tank air pump equipped with a multiple outlet

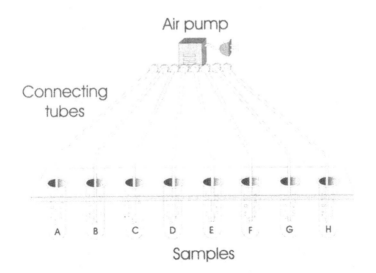

FIGURE 2. Experimental setup for the oxidation of Fe^{2+} ions by air at different pH values (A–H, arbitrary values). (Reprinted from The Chemical Educator, Vol. 9, No. 1, 2004, pp. 9–11, with permission).

(for example, of the type that distributes air to several fish tanks), and connect an outlet to each one of the separate test tubes containing the iron solutions. See Figure 2. Bubble air through all the solutions for 10 to 15 minutes. Then, filter approximately 2.5 mL with a 0.45 μm syringe membrane filter or a very fine filter paper. Take 2.0 mL from the resulting filtrate so as to analyze for the Fe(II) content by a standard potentiometric titration. To accomplish this, use 0.05 N $KMnO_4$ as the titrant and a commercial or a home-made oxidation–reduction potential electrode (ORP) as the sensor. Acidify each solution before titration with 0.5 mL of 3 M H_2SO_4. Use the microscale technique with a 2-mL serologic pipet.

Name_____Section_____Date_____

Instructor_____Partner_____

PRELABORATORY REPORT SHEET—EXPERIMENT 7

Experiment Title: _____

Objectives

Flow sheet of procedure

Waste containment and recycling procedure

At the end of the experiment, return all portions of the Fe(II)/Fe(III) solutions to the instructor.

PRELABORATORY QUESTIONS AND PROBLEMS

1. Write the balanced equations for the formation of colored complexes of iron:

a) $Fe^{2+} + [Fe(CN)_6]^{3-} \rightarrow$
b) $Fe^{3+} + [Fe(CN)_6]^{4-} \rightarrow$
c) $Fe^{3+} + SCN^- \rightarrow$

2. Write the balanced equation for the oxidation of Fe(II) by potassium permanganate.

3. Write down the necessary equations to calculate the Fe(II) concentration knowing the volume of $KMnO_4$ titrant and its concentration.

Additional Related Projects

• The potentiometric $KMnO_4$ titration can be substituted by a titration with periodate. (See Drummond, 1997).

• The experiment can be carried out at different temperatures and different dioxygen concentrations.

Name_____Section_____Date_____

Instructor_____Partner_____

LABORATORY REPORT SHEET—EXPERIMENT 7

Observations

Part I

Result of test for Fe^{3+} ions:

Test tube I   positive _____   negative _____

Test tube II   positive _____   negative _____

Part II

Titration results

Time of air bubbling: _____min

Sample 1:

Initial pH: _____

Volume of aliquot: _____mL

Vol. titrant, mL	Reading on potentiometer
0	

Equivalence point: _____mV

[Fe^{2+}]: _____M

[Fe^{3+}]: _____M

[Fe^{2+}]/[Fe^{3+}]: _____

Sample 2:

Initial pH: _____

Volume of aliquot: _____mL

Vol. titrant, mL	Reading on potentiometer
0	

Equivalence point: _____mV

[Fe^{2+}]: _____M

[Fe^{3+}]: _____M

[Fe^{2+}]/[Fe^{3+}]: _____

Sample 3:

Initial pH: _____

Volume of aliquot: _____mL

Vol. titrant, mL	Reading on potentiometer
0	

Equivalence point: _____mV

$[Fe^{2+}]$: _____M

$[Fe^{3+}]$: _____M

$[Fe^{2+}]/[Fe^{3+}]$: _____

Sample 4:

Initial pH: _____

Volume of aliquot: _____mL

Vol. titrant, mL	Reading on potentiometer
0	

Equivalence point: _____mV

$[Fe^{2+}]$: _____M

$[Fe^{3+}]$: _____M

$[Fe^{2+}]/[Fe^{3+}]$: _____

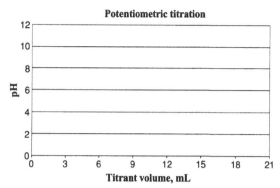

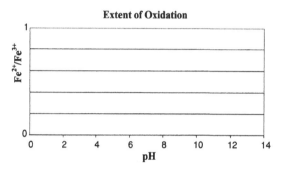

POSTLABORATORY PROBLEMS AND QUESTIONS

1. From the results obtained in Part 1, analyze the relationship of the pH with the oxidation of Fe(II).

2. Plot at least three of the potentiometric titrations and obtain the equivalence points.

3. Plot the Fe(II)/Fe(III) ratio vs pH and conclude how the extent of oxidation of Fe(II) depends on pH.

4. Discuss how the change from air to pure oxygen could affect the extent of oxidation of Fe(II).

Student Comments and Suggestions

Literature References

Drummond, T. G.; Lockhart, W. L.; Slattery, S. J.; Khan, F. A.; Leavitt, A. J. "Periodate Titration of Fe(II) in

Acid Aqueous Solutions: An Environmentally Friendly Redox Reaction for the Undergraduate Laboratory," *Chem. Educ.* **1997**, *2(4)* 1–7.

Emmenegger, L.; King, D. W.; Sigg, L.; Sulzberger, B. "Oxidation Kinetics of Fe(II) in a Eutrophic Swiss Lake," *Environ. Sci. Technol.* **1998**, *32*, 2990–2996.

Fregoso-Infante, A.; Ibanez, J.G.; Gonzalez-Rosas, L.C.; Garcia-Pintor, E. "Microscale Environmental Chemistry. Part 3. The Oxidation of Fe(II) by Molecular Oxygen as a Function of pH," *Chem. Educator* **2004**, *9*, 9–11.

Ibanez, J. G.; Miranda, C.; Topete, J.; García, E. "Metal Complexes and the Environment: Microscale Experiments with Iron-EDTA Chelates," *Chem. Educator* **2000**, *5*, 226–230.

Ibanez, J. G.; Hernandez-Esparza, M.; Valdovinos-Rodriguez, I.; Lozano-Cusi, M.; de Pablos-Miranda, A. "Microscale Environmental Chemistry. Part 2. Effect of Hydrogen Peroxide in the Presence of Iron (II) (Fenton Reagent) and Other Conditions upon an Organic Pollutant," *Chem. Educator* **2003**, *8*, 47–50.

King, D. W. "Role of Carbonate Speciation on the Oxidation Rate of Fe(II) in Aquatic Systems," *Environ. Sci. Technol.* **1998**, *32*, 2997–3003.

Kotronarou, A.; Sigg, L. "Sulfur Dioxide Oxidation in Atmospheric Water: Role of Iron(II) and Effect of Ligands," *Environ. Sci. Technol.* **1993**, *27*, 2725–2735.

Pang, H.; Zhang, T. C. "Fabrication of Redox Potential Microelectrodes for Studies in Vegetated Soils or Biofilm Systems," *Environ. Sci. Technol.* **1998**, *32*, 2990–2996.

Rajeshwar, K.; Ibanez, J. G.; *Environmental Electrochemistry: Fundamentals and Applications in Pollution Abatement*; Academic Press: San Diego, CA, 1997. Chapters 1, 5, 8.

Richardson, J. N.; Stauffer, M. T.; Henry, J. L. "Microscale Quantitative Analysis of Hard Water Samples Using an Indirect Potassium Permanganate Redox Titration," *J. Chem. Educ.* **2003**, *80*, 65–67.

Schnell, S.; Ratering, S.; Jansen, K.-H. "Simultaneous Determination of Iron(III), Iron(II), and Manganese(II) in Environmental Samples by Ion Chromatography," *Environ. Sci. Technol.* **1998**, *32*, 1530–1537.

Singh, M. M.; McGowan, C.; Szafran, Z.; Pike, R. M. "A Modified Microburet for Microscale Titration," *J. Chem. Educ.* **1998**, *75*, 371.

To, T. B.; Nordstrom, D. K.; Cunningham, K. M.; Ball, J. W.; McCleskey, R. B. "New Method for the Direct Determination of Dissolved Fe(III) Concentration in Acid Mine Waters," *Environ. Sci. Technol.* **1999**, *33*, 807–813.

Viswanathan, M. N.; Boettcher, B. "Biological Iron Removal from Groundwater: A Review," *Wat. Sci. Technol.* **1991**, *23*, 1437–1446.

Experiment 8
Photoassisted Reduction of Metal Complexes

Reference Chapters: 6, 8, 10

Objectives

After performing this experiment, the student shall be able to:

- Understand what a metal chelate is and prepare one in the laboratory.
- Explain the effect of light on the rupture of a metal-complex bond.
- Produce and identify a metal ion from the photolysis of a metal chelate.

Introduction

Ligands are present in the environment due to a plethora of natural and anthropogenic events. They affect metal adsorption onto natural oxides, the dissolution of natural oxides and of metal scales, metal ion removal and uptake by biological systems, metal-solubilization in soils, and the like. Thus, an understanding of the environmental reactions and fate of ligands and metal complexes is of keen importance.

As discussed in Chapters 2 and 6, natural light induces many processes of environmental significance either directly (when light is absorbed by the species of interest) or indirectly (when light excites or produces an intermediate species called *mediator* that then reacts with the species of interest). For example, the photolysis of metal complexes is important because it obviously affects their concentration and fate (see Section 6.3). A typical example is the photolysis of iron—carboxylato complexes (i.e., complexes with oxalate, aminopolycarboxylates, citrate and humic and fulvic acids) that can occur with high quantum yields. For this reason, in the present experiment an FeEDTA complex is formed and photolyzed, and its metal ion product made evident by a color-producing reaction. EDTA was selected because it is a strong complexing agent for many metal ions and a persistent pollutant due to its high stability (it is even resistant to decomposition by ionizing radiation and heat) and low biodegradability.

Photodegradation of [Fe(III)EDTA]

As discussed in Section 6.3 some EDTA complexes are light-sensitive; for instance, [Fe(III)EDTA] can undergo total photolysis in a sunny day within several hours. Others are only slightly affected (e.g., [Mn(II)EDTA], [Co(III)EDTA]), while others are not affected by light at all. The ability of [Fe(III)EDTA] to undergo a photoredox reaction is very fortunate because—as stated above—EDTA is a refractory compound and thus the natural photolytic pathway provides a means for its destruction.

A simplified photoreduction reaction using a carboxylic anion as the ligand (e.g., oxalate) is the following:

$$[Fe(III)(C_2O_4)_3]^{3-} + h\nu$$
$$\rightarrow Fe(II) + 2C_2O_4{}^{2-} + CO_2{}^{-\cdot} + CO_2 \quad (1)$$

Note that in the $C_2O_4{}^{2-}$ ligand, C has an oxidation state of 3+ whereas in CO_2 it is 4+. Then, C underwent oxidation in order to reduce Fe(III) to Fe(II), which is then released from the complex.

In the following qualitative experiment, an [Fe(III)EDTA] complex will be exposed to light (either natural or artificial) and decomposed to produce Fe(II), which then reacts with iron hexacyanate (III) yielding a highly colored solution of another complex (see Ibanez, 2000).

Experimental Procedure

Estimated time to complete the experiment: 1 h

Materials	Reagents
3 10-mL volumetric flasks	$FeCl_3 \cdot 6H_2O$
1 1-mL graduated pipet	$K_3[Fe(CN)_6]$
3 Beral pipets	Na_2H_2EDTA
5 test tubes, 5 cm-long	6 M NaOH
1 Slide projector	D. I. water
Aluminum foil	

Prepare 10 mL of an Fe(III) solution by adding 0.8 g of $FeCl_3 \cdot 6H_2O$ to 8 mL of H_2O in a volumetric flask. Stir and dilute to the 10-mL mark. In another 10-mL flask dissolve 1 g of $K_3[Fe(CN)_6]$ and dilute to the mark. Lastly, add 1 g of Na_2EDTA to 5–6 mL of H_2O in a third 10-mL flask, and add approximately 1 mL of 6 M NaOH to facilitate dissolution of the EDTA species; then, dilute to the mark. These procedures yield three solutions (approximately 0.3 M each) that should be enough for a large class because only one drop of each is needed. **Caution: K_3 $(Fe(CN)_6)$ is harmful by inhalation, ingestion and through skin contact**.

Now, place approximately 1 mL of distilled water, one drop of the $FeCl_3$ solution, and one drop of the $K_3[Fe(CN)_6]$ solution in a 5-cm long test tube. Do this in duplicate (the second one is the *blank* test tube). Add one drop of the Na_2 EDTA solution to each tube and shake. Wrap the blank test tube with aluminum foil so as to keep it away from light. Expose the other test tube to sunlight or to the light of an overhead projector (or of a slide projector). In a few minutes, a dramatic color change is observed. The blank test tube is then unwrapped, and the difference between both is annotated.

Name_____Section_____Date_____

Instructor_____Partner_____

PRELABORATORY REPORT SHEET—EXPERIMENT 8

Objectives

Flow sheet of procedure

Waste containment and recycling procedure

PRELABORATORY QUESTIONS AND PROBLEMS

*1. The pH is a key parameter in the present experiment. Calculate and plot the distribution diagram of EDTA as a function of pH. Predict what would happen if a pH of 2 were used? A pH of 4? Why is a high pH recommended for the present experiment?

2. Find in the literature at least three direct photolytic processes used for the destruction of wastes.

3. When should an indirect method be considered for the photodestruction of a waste?

_____

* Answer in this book's webpage at www.springer.com

Additional Related Projects

- Study the photolysis kinetics in this experiment by taking the absorbance of the complex obtained as a function of time.
- Analyze the effect of pH in the present experiment by using different values.
- Design an experiment to demonstrate the production of CO_2 during the photolysis of the FeEDTA used in this experiment. (See Lockhart and Blakeley, 1975).
- Design and perform experiments with other ligands (e.g., citrate, oxalate, some non-carboxylated organics, etc.) and other iron (II/III) indicators (e.g., o-phenanthroline, $K_4[Fe(CN)_6]$, SCN^-).
- Try to photolyze a Mn(II)EDTA or Co(III)EDTA complex in the same manner as it is done with Fe(II)EDTA in the present experiment. Perform qualitative tests for the presence of the metal ions after photolysis.

Name_____Section_____Date_____

Instructor_____Partner_____

LABORATORY REPORT SHEET—EXPERIMENT 8

Observations

1. Color of the initial reaction mixture   _____

2. Color of the reaction mixture after irradiation   _____

3. Color of the blank after irradiation   _____

4. Time allowed for the reaction   _____ min

POSTLABORATORY PROBLEMS AND QUESTIONS

*1. Discuss the reason for the color change in one tube and not in the other.

*2. If Fe(III) were photo produced instead of Fe(II), what indicator would you use?

*3. The $C_2O_4^{-\cdot}$ radical is a precursor for the $CO_2^{-\cdot}$ product from reaction 1, and can react with ambient dioxygen to yield the radical anion X (that contains only O). In acidic media, the diatomic X reaches an equilibrium with a triatomic neutral radical, Y (that contains H and O). X and Y disproportionate in acid to give dioxygen and a neutral species, Z that is capable of transforming Fe(II) back to Fe(III). Z contains H and O, possesses an O-O bond, and can undergo disproportionation (dismutation) to yield dioxygen plus water. What are the chemical formulas of X, Y, and Z?

_____
* Answer in this book's webpage at www.springer.com

Student Comments and Suggestions

Literature References

Balmer, M. E.; Sulzberger, B. "Atrazine Degradation in Irradiated Iron/Oxalate Systems: Effects of pH and Oxalate," *Environ. Sci. Technol.* **1999**, *33*, 2418–2424.

Hislop, K. A.; Bolton, J. R. "The Photochemical Generation of Hydroxyl Radicals in the UV-vis/Ferrioxalate/H₂O₂ System," *Environ. Sci. Technol.* **1999**, *33*, 3119–3126.

Hug, S. J.; Laubscher, H.-U. "Iron (III) Catalyzed Photochemical Reduction of Chromium (VI) by Oxalate and Citrate in Aqueous Solutions," *Environ. Sci. Technol.* **1997**, *31*, 160–170.

Ibanez, J. G. "Redox Chemistry and the Aquatic Environment: Examples and Microscale Experiments," *Chem. Educ. Int. (IUPAC)* **2005**, *6*, 1–7.

Ibanez, J. G.; Miranda-Treviño, J. C.; Topete-Pastor, J.; Garcia-Pintor, E. "Metal Complexes and the Environment: Microscale Experiments with Iron–EDTA Chelates," *Chem. Educ.* **2000**, *5*, 226–230.

Llorens-Molina, J. A. "Photochemical Reduction of Fe^{3+} by Citrate Ion," *J. Chem. Educ.* **1988**, *65*, 1090.

Lockhart, H. B., Jr.; Blakeley, R. V. "Aerobic Photodegradation of Fe(III)-(Ethylenedinitrilo)tetraacetate (Ferric EDTA). Implications for Natural Waters," *Environ. Sci. Technol.* **1975**, *12*, 1035–1038.

McArdell, C. S.; Stone, A. T.; Tian, J. "Reaction of EDTA and Related Aminocarboxylate Chelating Agents with $Co^{III}OOH$ (Heterogenite) and $Mn^{III}OOH$ (Manganite)," *Environ. Sci. Technol.* **1998**, *32*, 2923–2930.

Porter, G. B. "Introduction to Inorganic Photochemistry: Principles and Methods," *J. Chem. Educ.* **1983**, *60*, 785–790.

Porterfiel, W. "Photochemical Reactions of Transition Metals," Chapter 17 in: *Inorganic Chemistry: A Unified Approach*; Academic Press, New York, 1993.

Stumm, W.; Morgan, J. J. *Aquatic Chemistry: Chemical Equilibria and Rates in Natural Waters*; Wiley Interscience: New York, 1996.

Tiemann, K. J.; Gardea-Torresdey, J. L.; Gamez, G.; Dokken, K.; Sias, S. "Use of X-Ray Absorption Spectroscopy and Esterification to Investigate Cr(III) and Ni(II) Ligands in Alfalfa Biomass," *Environ. Sci. Technol.* **1999**, *33*, 150–154.

Xue, Y.; Traina, S. J. "Oxidation Kinetics of Co (II)-EDTA in Aqueous and Semi-Aqueous Goethite Suspensions," *Environ. Sci. Technol.* **1996**, *30*, 1975–1981.

Experiment 9
Anionic Detergents and *o*-Phosphates in Water

Reference Chapters: 6, 8

Objectives

After performing this experiment, the student shall be able to:

- Determine the concentration of anionic detergents in a sample of water using the methylene blue-active substances technique.
- Determine the presence of soluble orthophosphates by the colorimetric molybdate method.

Introduction

Detergents are among the most common water pollutants, discharged as a consequence of laundering or cleaning processes, either from households or industrial origin. Detergents are mainly preparations or mixtures of linear alkyl sulfonates and other additives that help remove grease and dirt. Besides being slowly biodegradable, they are mainly surfactants that have the property of emulsifying grease and dirt, and forming foams. These compounds and their foams are inconvenient in water discharges because they interfere with the transfer of air to water. In addition, they may deflocculate colloids, promote the flotation of solids, emulsify oil and grease, lower the level of dissolved oxygen through biodegradation, and have a negative aesthetic impact. They can also destroy the natural water-repellent protective coating of aquatic animals and birds, which may

- facilitate parasite attack
- unbalance their water exchange
- produce sliming of gills in fish

In large concentrations, detergents may cause the death of aquatic plants and animals.

The majority of the surfactants used in detergents are of anionic nature, linear or non-linear with sulfonate ($R\text{-}SO_3^-$) or sulfate ester groups ($R\text{-}OSO_3^-$). The analytical method used in this experiment to determine their concentration in water involves the reaction of these anionic compounds with a cationic dye (methylene blue), which forms a strong ion-pair that can be separated from the aqueous phase with an organic solvent. (This is the basis of the USEPA method 425.1 and of the ASTM D2330 method). The intensity of the blue color in the extracted phase relates to the amount of anionic surfactant present and can be measured spectrophotometrically at 650 nm. With this method, detergent concentrations in the $\mu g/L$ to mg/L can be detected in water or in wastewater. Substances that react with methylene blue are called *methylene-blue active substances (MBAS)*. Although several interferences can affect this method, they are rarely found in normal samples. In this sense, the only strong interference for this method is a very high concentration of chloride ions (> 1000 mg/L).

Phosphates were profusely employed earlier as additives in the preparation of detergents and cleaning formulas. They are also produced as a consequence of

- soil erosion processes
- runoff containing fertilizers
- boiler water discharges (that use phosphates as additives)
- biodegradation of organic matter from urban discharges

Phosphates are typically found in water as ortho-phosphates, but there are also other more complex forms that are hydrolyzable condensed inorganic phosphates such as meta-, pyro-, and polyphosphates, and the organically bound phosphorous. The sum of these forms (either suspended or dissolved) is known as the *total phosphorous content* of water. The dissolved inorganic form is the one generally measured.

Excessive amounts of phosphorous, together with nitrogen sources derived from discharges, cause a proliferation of aquatic plants and algae known as *accelerated* or *cultural eutrophication* that may ruin the quality of lakes and ponds. A mere 0.005 mg/L of inorganic phosphorous above normal levels is enough for this process to begin. With the proliferation of detergents, the phosphorous content in surface water has doubled (or even tripled in some cases). In view of this, the use of phosphate as an additive in detergents has been greatly reduced in most developed countries.

One of the analytical methods for phosphate determination consists in the reaction of a water sample with ammonium molybdate under acidic pH conditions. This generates phosphomolybdic acid, which is then reduced with tin (II) chloride, ascorbic acid, or an amino acid to form a blue molybdenum complex. The intensity of the color of the complex developed is proportional to the concentration of orthophosphate present and can be quantified colorimetrically at 690 nm with the aid of a calibration curve. If one desires to measure the condensed inorganic form, it must be previously converted into orthophosphate by an acid hydrolysis. It is common that part of the organic phosphorous will be released with this process; however, if one needs to know the total organic phosphorous content, then a strong acid oxidation must be carried out using persulfate in a strongly acidic medium. With the hydrolyzed or oxidized samples, the orthophosphate colorimetric technique is followed. Three different values are then obtained: (A) for the direct orthophosphate measurement, (B) for the acid hydrolysis conversion of the inorganic and partial organic phosphorous, and (C) for the strong oxidative digestion of the organic phosphorous. Then, to determine the total acid-hydrolyzable amount of phosphorous (inorganic and organic) one must subtract A from B, (B−A); and for the amount of organic phosphorous content, one must subtract B from C, (C−B).

In this experiment a synthetic sample will be prepared by adding a (small) known amount of detergent to tap water or to D.I. water. Alternatively, one can obtain a water sample from a laundromat or laundry system discharge.

The experiment can also be done with two different types of detergents: one that contains phosphates and one without phosphates. Because the methylene blue method is only applicable to anionic detergents, it can be interesting for the students to observe the response of a water sample containing soap. Another sample that can be tested is from a surface water system where a wastewater treatment plant discharges.

The experiment consists in preparing a calibration curve for a linear alkyl sulfonate (LAS), such as lauryl sodium sulfonate or lauryl ammonium sulfate, and a phosphorous calibration curve (as *o*-phosphate) with the two techniques mentioned above. Finally, the spectrophotometric response of each sample will be measured.

Experimental Procedure

Estimated time to complete the experiment: See each individual procedure. This experiment may be completed in one or two lab sessions.

Materials	Reagents
2 2-mL volumetric pipets	0.001 M and 0.01 M HCl
4 2-mL graduated pipets	0.01 M Na_2CO_3
2 1-mL graduated pipets	0.01 M NaOH
2 5-mL graduated pipets	6 N H_2SO_4
6 10-mL beakers	concentrated H_2SO_4
4 Beral pipets	$NaH_2PO_4 \cdot H_2O$
10 10-mL volumetric flasks	Phenolphthalein indicator
2 100-mL volumetric flasks	Methylene chloride
2 propipets	**Methylene blue complexing**
4 long tip Pasteur pipets	**solution:**
with small latex bulbs	(Dilute 100 mg of methylene
6 10-mL capped vials or	blue in 100 mL of D.I.
test tubes	water. Transfer 3 mL of the
2 Pasteur pipets packed	solution to a 100-mL
with glass fiber or cotton	volumetric flask and
filter and with small latex	half-fill it with D.I. water.
bulbs	Then, add 4 mL of 6 N
1 test tube rack	sulfuric acid and 5 g of
1 D.I. wash bottle	$NaH_2PO_4 \cdot H_2O$. Mix well
1 VIS-spectrophotometer	until dissolved and add D.I.
3 spectrophotometer glass	water as needed to reach the
or quartz cells	100-mL mark).
1 pH meter with a test tube	**Washing solution**:
pH probe	(In a 100-mL volumetric
	flask add 0.7 mL of
	(cont.)

Materials	Reagents
	concentrated sulfuric acid and dilute it with 50 mL of D.I. water. Add 5 g of $NaH_2PO_4 \cdot H_2O$ and mix until dissolved. Then fill up to the 100-mL mark of the flask with D.I. water). **Standard solution of LAS:** (Prepare a 100 mg/L solution of either ammonium lauryl sulfate, the sodium salt of dodecylbenzensulfonic acid, sodium lauryl sulfate, or any other linear anionic surfactant).

Safety Measures

Strict precautions must be observed in the handling of the organic solvent (i.e., methylene chloride) because it is carcinogenic. It is therefore important to prevent its volatilization as much as possible. Handle this compound carefully under an extraction hood, preventing its inhalation and dermal contact. Although the amounts used are relatively small, there is always risk of vaporization and spills. One must also avoid contact of any acids (used for adjusting pH) with the skin or eyes, because they are corrosive. All the residues generated in this experiment must be disposed in special bottles identified for that purpose.

A. Measurement of the MBAS content in water samples

Estimated time required: 5–10 min per sample

Samples

Take a sample:

a) From a surface water system: lake, pond or river
b) From the discharge of a washing machine, or prepare a synthetic sample adding a very small amount of detergent in water (check whether the list of contents mentions a phosphate additive or not). If possible, have a sample with phosphate and another one without it.
c) From the output of a wastewater treatment plant. **(Note: In this last case, extreme safety measures must be taken in handling the samples because they will probably contain pathogens.**

Use gloves, safety goggles, and protective clothing, and be sure to disinfect the lab working zone after the experiment).

d) Prepare another sample with soap so as to have students visualize that these compounds do not form the ion-pair with the cationic methylene blue and thus will not respond to the test.

A.1 Dilutions of the LAS standard solution for the calibration curve.

In 10-mL volumetric flasks, prepare the following volumes (in mL) of the LAS standard solution (100 mg/L) and D.I. water, respectively: 4/10, 5/10, and 6/10.

Then, follow the technique in section **A.2** to measure their LAS content. Using this same technique, prepare a blank (i.e., D.I. water).

The detection limit for MBAS using this technique is 1 mg/L of LAS.

A.2 Technique

The following technique applies to all the water samples and to the solutions prepared with the LAS standard solution so as to prepare a calibration curve.

Experimental steps

1. Measure the pH of each sample or solution. If it is outside the pH range from 8.0 to 8.4, take a 2-mL aliquot and place it in a vial or in a capped test tube. Add one or two drops of phenolphthalein indicator and look for any color development. If colorless, add a sodium carbonate or sodium hydroxide solution drop wise until the endpoint is reached (i.e., a pale pink color); then, add diluted HCl acid drop wise until the pink color disappears.

2. If the initial pH is greater that 8 (which seldom happens), place a 2-mL aliquot in a vial and add one or two drops of phenolphthalein indicator. Upon addition of the phenolphthalein, the pink color will appear. Add enough diluted HCl acid drop wise until the color disappears.

3. Adjust the pH of each sample to 8.3 and then add 0.5 mL of the methylene blue complexing solution and 2 mL of methylene chloride. **(Caution: Work under a fume or extraction hood.)** Close the cap immediately and tightly, and mix softly. Note what happens to the reaction mixture. Identify the organic and aqueous phases.

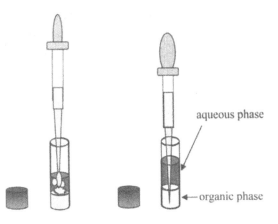

FIGURE 1. Organic phase mixing and extraction.

4. Shake the reaction mixture for 20 s and then slightly loosen the vial's cap to liberate the pressure of the solvent generated by the mixing. **Caution: Work under an extraction hood and aim the mouth of the vial away from you or any other person nearby**.
A mixing alternative consists in extracting the lower phase with a Pasteur pipet and allowing it to rapidly fall with the aid of some pressure through the aqueous phase (so as to generate turbulence). Repeat the procedure approximately 10 times (see Figure 1).

5. Cap the vial and allow the contents to stand still until the two phases separate. Note whether there is any change in the color of the organic phase and its intensity. Does your sample have a large amount of anionic detergent?

6. Separate the organic phase from the aqueous phase using a long-tip Pasteur pipet with a rubber bulb. To prevent any undesired transference of the aqueous phase with the organic phase, introduce the tip of the pipet all the way down to the bottom of the vial while pressing the rubber bulb, and carefully suction liquid into the pipet until all the organic phase is removed. Transfer it directly into a new capped vial (see Figure 1).

Repeat this extraction procedure from the aqueous phase twice with 2 mL of organic solvent each time, and in the same way transfer it to the capped vial where the first extract was deposited. Add 5 mL of washing solution into this vial, and shake or mix thoroughly for 30 s (or 10 times). Allow it to rest until the phases separate, and then extract the organic phase again. Only if the organic phase were turbid, filter it through a cotton or glass wool filter. Receive the organic phase free of suspended solids in a 10-mL volumetric flask (see Figure 2).

Add 2 mL of solvent to the remaining washing solution and perform an extraction. Transfer the extract to the volumetric flask, and add solvent up to the 10-mL mark, if needed. **Caution: Always cap the systems containing the organic solvent to avoid evaporation, and work inside the extraction hood**.

With the extracts contained in the volumetric flask, fill a spectrophotometer cell (glass or quartz) and measure the absorbance at 650 nm. Make sure that the spectrophotometer is previously calibrated using the methylene chloride so as to adjust the transmittance to 100%.

This technique will be used with each one of the samples and with the dilutions of the standard LAS solution. With the absorbance values of these last solutions and their real concentrations, generate the calibration curve and its corresponding equation.

B. Measurement of the orthophosphate content in water samples.

Estimated time required: 15 min per sample

Safety measures

Take the common precautions in handling acids and bases and the molybdate reagent. All the residues

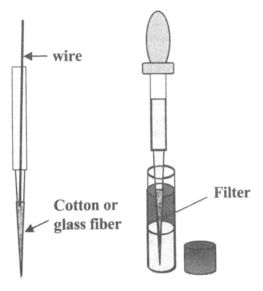

FIGURE 2. Filter preparation and organic phase filtration.

generated in this experiment must be disposed in special bottles identified for that purpose.

Materials	Reagents
3 1-mL (1/100) graduated pipets	**P-as phosphate standard solution,**
1 2-mL (1/100) graduated pipet	10 mg/L (dissolve 0.04 g of KH$_2$PO$_4$ in 1 L of water)
1 5-mL graduated pipet	0.01 N HCl solution
4 10-mL beakers	Phenolphthalein indicator
3 propipets	0.1 M Na$_2$CO$_3$ or 0.01 N NaOH for pH adjustment
4 Beral pipets	
10 25-mL Erlenmeyer flasks	**Stannous chloride reagent** (dissolve 0.25 g of bi-hydrated tin (II) chloride in 10 mL of glycerol; warm in a water bath with mixing until dissolution and allow it to cool)
12 10-mL volumetric flasks	
1 D.I. water wash bottle	
1 chronometer	
1 portable filter holder	
1 5-mL syringe	
0.8 μm acetate filter membranes	**Ammonium molybdate reagent** (dissolve 2.5 g of tetrahydrated ammonium molybdate in 20 mL of D.I. water in a 100-mL volumetric flask. Dissolve 28 mL of concentrated H$_2$SO$_4$ in 40 mL of D.I. water, allow it to cool, and add to the ammonium molybdate solution. Then, add D.I. water up to the 100-mL mark)
1 bottle for residues	
1 spectrophotometer	
3 spectrophotometer glass cells	
1 pH meter with a test tube pH probe	

B.1 Dilutions of the standard o-phosphate solution for the calibration curve.

In 10-mL volumetric flasks prepare the following mL of the standard solution (10 mg/L) and D.I. water: 0.2/10; 0.5/10; 1.0/10; 1.5/10; 2.0/10; 2.5/10; 3.0/10; 4/10; and 5/10.

After preparing these solutions follow the technique in section **B.2** to measure the amount of *o*-phosphate present. Also prepare a blank of D.I. water by following the same technique.

The detection limit for the *o*-phosphate measurement with this technique is 0.5 mg/L of P.

B.2 Technique

The following technique applies to the water samples as well as to the standard dilutions. Note: The sample must be free of suspended solids before applying the technique; therefore, the water samples must be previously filtered through a membrane filter.

1. With a 1-mL graduated pipet (1/100) take 0.5 mL of sample and place in a 10-mL volumetric flask; fill to the 10-mL mark with D.I. water and mix perfectly.

2. Transfer the contents of the volumetric flash to a 25-mL Erlenmeyer flask and add phenolphthalein indicator drop wise. Observe if there is any color formation.

   If a pink color appears, add diluted hydrochloric acid drop wise until the endpoint is reached. If no pink color appears, add some drops of a base solution (i.e., of Na$_2$CO$_3$ or dilute NaOH) until the pink color appears. Then, add 0.01 M HC1 drop wise until the phenolphthalein endpoint is reached. What is the reason for performing this adjustment?

3. To the sample in the Erlenmeyer flask, add 0.4 mL of the ammonium molybdate reagent with a 1-mL graduated pipet. Mix, and then add 2 drops of the tin (II) chloride solution. Mix and allow it to react for 10 minutes. Note all the changes.

4. At the end of this reaction time, if there were *o*-phosphate present, a blue-colored solution will be observed and the color intensity will be proportional to the concentration. Once the reaction is over (and before 12 min), measure the absorbance of the treated sample in the spectrophotometer at a wavelength of 690 nm, previous adjustment to 100% transmittance with D.I. water. If the sample is measured later, some turbidity and a change in color will be observed, and accuracy will be accordingly lost.

   If any of the water samples prepared with this technique develops a very intense or dark blue color, any measurement will fall outside the range of the calibration plot. In this case, repeat the preparation of the sample but dilute it to a known value (e.g., 50%, or 1:2). Then, apply the technique and correct the measured concentration by the dilution factor.

5. With the absorbance values of the diluted standard, calculate the real concentrations, build the calibration curve, and find its corresponding equation for a linear fit. This will allow the calculation of the real *o*-phosphate concentration in each one of the samples.

   Apply this technique to each sample and record the absorbance values.

Name_____Section_____Date_____

Instructor_____Partner_____

PRELABORATORY REPORT SHEET—EXPERIMENT 9

Objectives:

Flow sheet of procedure:

Waste containment procedure:

PRELABORATORY QUESTIONS AND PROBLEMS

1. Explain why is it important to measure the MBAS content of an environmental sample of water. What are the negative effects of the presence of these compounds in aquatic natural systems?

2. Explain why is it important to measure the phosphorus or phosphate content of an environmental sample of water. What are the positive and negative effects of the presence of such ions in aquatic natural systems?

3. Seek in the literature the maximum permissible concentrations in drinking water for anionic surfactants and for total phosphate or phosphorus content.

4. What are the limits of phosphorus concentration in water in order to have a eutrophic lake or river?

5. Establish the reactions involved in the phosphate analysis, including the one related to the acid hydrolysis of the condensed inorganic phosphorous.

6. Give two reasons as of why the methylene blue method cannot be used for cationic surfactants.

7. What would it happen if the methylene blue method were applied to a soap-containing water sample?

8. What are the main interferences in the two methods studied in this experiment?

9. For the MBAS method, what is the reason for the acid wash of the organic phase (once the extraction has been carried out)?

Additional Related Projects

• Measure the total inorganic phosphate and the condensed or hydrolyzable phosphates by skipping the filtration step described above. This is achieved by the transformation of the meta-, pyro-, and polyphosphates into orthophosphates (through acid hydrolysis); then, obtain the total inorganic phosphate content as well as the fraction that corresponds to the condensed forms. This may be performed with a 10-mL sample treated with 0.2 mL of concentrated sulfuric acid to ensure an acidic pH. The mixture is then heated and allowed to boil gently for a minimum of 5 minutes. Allow it to cool, adjust the volume again to 10 mL with D.I. water, and apply the *o*-phosphate analytical technique described earlier.

- A similar procedure determines the total phosphorous content. In this case, treat a 10-mL sample with 0.5 mL of concentrated sulfuric acid so that the pH is < 1; then add 2 mL of a 5% potassium persulfate solution and bring the resulting mixture to boil for at least 30 minutes. Allow it to cool, adjust the volume again to 10 mL with D.I. water, and apply the *o*-phosphate analytical technique described earlier.

Name_____Section_____Date_____

Instructor_____Partner_____

LABORATORY REPORT SHEET—EXPERIMENT 9

PART A. *Measurement of the anionic surfactant content in water samples.*

Experimental data

A.1 LAS calibration curve.

Anionic surfactant used as standard:_____

Concentration of the standard solution:_____mg/L of LAS

Wavelength used in the spectrophotometric measurement:_____nm

Volume of standard/ 10 mL of solution	Concentration of anionic surfactant, mg/L	Absorbance
0		
0.3		
0.5		
0.8		
1		
1.5		
2		
2.5		
3		
3.5		
4		

Calibration curve equation: mg/L LAS =

Correlation coefficient, r^2 =

A.2 Surfactant content in several water samples.

Sample A.2.1 Surface water.

Source:_____

Observable characteristics:_____

_____

pH of original sample:_____

Sample A.2.2 Detergent-containing water

*Source:*_____

*Observable characteristics:*_____

_____

*pH of original sample:*_____

Sample A.2.3 Detergent-containing water

*Source:*_____

*Observable characteristics:*_____

_____

*pH of original sample:*_____

Sample A.2.4 Soap-containing water

*Source:*_____

*Observable characteristics:*_____

_____

*pH of original sample:*_____

Sample A.2.5 Wastewater treatment plant sample

*Source:*_____

*Observable characteristics:*_____

_____

*pH of original sample:*_____

Sample No.	Type of sample	Absorbance	LAS concentration, mg/L
A.2.1			
A.2.2			
A.2.3			
A.2.4			
A.2.5			

Calculation example:

PART B. Measurement of the *o*-phosphate content in water samples.

Experimental data.

B.1 P-phosphate calibration curve.

*Phosphate compound used as standard:*_____

*Concentration of the standard solution:*_____*mg/L of phosphorus (P)*

:_____*mg/L of phosphate* ion (PO_4^{3-})

*Wavelength used in the spectrophotometric measurement:*_____*nm*

Volume of standard/10 mL of solution	Concentration of P, mg/L	Absorbance
0		
0.2		
0.5		
1		
1.5		
2		
2.5		
3		
4		
5		

Calibration curve equation: mg/L LAS = _____

Correlation coefficient, r^2 = _____

B.2 P-o-phosphate content in several water samples.

Sample B.2.1 Surface water.

*Source:*_____

*pH of original sample:*_____

Sample B.2.2 Detergent-containing water

*Source:*_____

*pH of original sample:*_____

Sample B.2.3 Detergent-containing water

*Source:*_____

*pH of original sample:*_____

Sample B.2.4 Wastewater treatment plant sample

*Source:*_____

*pH of original sample:*_____

Sample B.2.5 Soap-containing water

*Source:*_____

*pH of original sample:*_____

Sample No.	Type of sample	Absorbance	P concentration, mg/L
B.2.1			
B.2.2			
B.2.3			
B.2.4			
B.2.5			

Calculation example:

POSTLABORATORY QUESTIONS AND PROBLEMS

1. According to your results of Part A, what differences do you observe among samples? Was this expected? Why?

2. In the sample B.2.1 (surface water), is the level of phosphate favorable for eutrophication? What other measurements should be carried out to have a firmer support for your answer?

3. From the other samples (B.2.2, B.2.3, B.2.4) what differences do you observe? Was this expected of such samples? Why?

4. With the sample corresponding to a wastewater treatment plant, are the levels above what is expected from a wastewater plant discharge?

5. What kind of chemical treatment would you propose for reducing the amount of phosphate coming out from a wastewater treatment plant? Write down the possible reactions.

Student Comments and Suggestions

Literature References

Sawyer, C. N.; McCarty, P. L; Parkin, G. F., *Chemistry for Environmental Engineering*; 5th ed.; McGraw Hill: New York, 2003.

Stumm, W.; Morgan, J. J., *Aquatic Chemistry: Chemical Equilibria and Rates in Natural Waters*; Wiley Interscience: New York, 1996.

Patnaik, P. *Handbook of Environmental Analysis: Chemical Pollutants in Air, Water, Soil and Solid Wastes*; CRC Lewis Publishers: Boca Raton, FL, 1997.

Experiment 10
Halogenated Hydrocarbons and the Ozone Layer Depletion

Reference Chapters: 4, 8

Objectives

After performing this experiment, the student shall be able to:

- Compare the reactivity of ozone for various physical and chemical conditions.
- Apply a volumetric technique for the detection and quantification of ozone.

Introduction

The Earth's atmosphere is composed of several layers: (a) the *troposphere* (the layer closest to the ground) where most of the weather occurs (such as rain, snow, and clouds), (b) the layer above the troposphere (the *stratosphere*), an important region in which effects such as the *ozone hole* and *global warming* originate. Supersonic jet airliners fly in the lower stratosphere (a historical example was the French *Concorde*), whereas subsonic commercial airliners are usually in the troposphere. The narrow region between these two parts of the atmosphere is called the *tropopause*.

Ozone forms a layer in the stratosphere, thinner in the tropics (around the equator) and denser toward the poles. The amount of ozone above a given point on the Earth's surface is measured in *Dobson units* (DU)—and is typically about 260 DU near the tropics and higher elsewhere, although there are large seasonal fluctuations. Ozone is produced when ultraviolet radiation—generated in the Sun—strikes the stratosphere, dissociating (or separating) dioxygen molecules (O_2) into atomic oxygen (O). Atomic oxygen quickly combines with more dioxy-

gen molecules to form ozone:

$$O_2 + h\nu \rightarrow O + O \tag{1}$$

$$O + O_2 \rightarrow O_3 \tag{2}$$

(wavelength, $\lambda \leq 240$ nm).

Up in the stratosphere, ozone absorbs some of the potentially harmful ultraviolet (UV) radiation from the Sun (i.e., at wavelengths between 240 and 320 nm) that can cause skin cancer and damage vegetation, among other effects.

Although the UV radiation dissociates the ozone molecule, ozone can reform through the following reactions, resulting in no net loss of ozone:

$$O_3 + h\nu \rightarrow O_2 + O \tag{3}$$

$$O + O_2 \rightarrow O_3 \tag{2}$$

Ozone is also destroyed by the following reaction:

$$O + O_3 \rightarrow O_2 + O_2 \tag{4}$$

The Chapman Reactions

The reactions 1 to 4 are known as the *Chapman reactions*. Reaction 2 becomes slower with increasing altitude, while reaction 3 becomes faster. The concentration of ozone is a balance between these competing reactions. In the upper atmosphere, atomic oxygen dominates where UV levels are high. Moving down through the stratosphere, UV absorption increases and ozone levels peak at roughly 20 km. As we move closer to the ground, UV levels decrease and ozone levels fluctuate (with a general decreasing trend). The layer of ozone formed in the stratosphere by these reactions is sometimes called the *Chapman layer.*

Molecular chlorine is easily photodissociated (i.e., split by sunlight):

$$Cl_2 + h\nu \rightarrow Cl + Cl \qquad (5)$$

This is the key to the timing of the ozone hole. During the polar winter, the cold temperatures that form in the "vortex" lead to the formation of polar stratospheric clouds. Heterogeneous reactions convert the reservoir forms of the ozone-destroying species (like chlorine and bromine) to their molecular forms. When the sunlight returns to the polar region during the spring in the southern hemisphere (corresponding to the northern hemisphere autumn), the Cl_2 is rapidly split into chlorine atoms, which lead to the sudden loss of ozone. This sequence of events has been confirmed by measurements before, during, and after the ozone hole.

There is still one more ingredient to consider in the broad picture of the ozone destruction. We still have most of the ozone, but we have not explained the chemical reactions in which atomic chlorine actually participates to destroy ozone. We discuss this next.

Catalytic Destruction of Ozone

Measurements of chemical species above the pole show high levels of active forms of chlorine. However, we still have many more atoms of ozone than we do of the active chlorine, so how it is possible to destroy nearly all of the ozone?

The answer to this question lies in what are known as *"catalytic cycles."* A catalytic cycle is one in which a molecule significantly changes or enables a reaction cycle without being altered by the cycle itself.

The production of active chlorine requires sunlight, and sunlight drives the following catalytic cycles thought to be the main cycles involving chlorine, responsible for destroying the ozone:

$$ClO + ClO + M \rightarrow Cl_2O_2 + M \qquad (6)$$
$$Cl_2O_2 + h\nu \rightarrow Cl + ClO_2 \qquad (7)$$
$$ClO_2 + M \rightarrow Cl + O_2 + M \qquad (8)$$

where M represents any mediator atom or molecule needed to absorb the excess energy of the intermediate formed.

Then:

$$2 \times (Cl + O_3) \rightarrow 2 \times (ClO + O_2) \qquad (9)$$

Net reaction:

$$2O_3 \rightarrow 3O_2 \qquad (10)$$

In this experiment we will explore the behavior of ozone toward different reaction conditions, such as the presence of UV light and of an organochlorinated compound.

Experimental Procedure

Materials	Reagents
4 plastic bags (polyethylene, 2 kg)	0.3 M KI
1 black plastic bag	0.0005 M $Na_2S_2O_3 \cdot 5H_2O$
2 10-mL syringes	CH_2Cl_2
1 1-mL volumetric pipet	0.5% starch solution
2 2-mL microburet	concentrated H_2SO_4
1 three-finger clamp	
1 magnetic bar	
1 magnetic stirrer	
2 10-mL beakers	
1 long wave UV pencil lamp	
1 stop watch	
3 10-mL Erlenmeyer flasks	
1 propipet bulb	
1 universal stand	
1 ozone generator	

Experiments

I) Ozone vs time

Connect the exit of the ozone generator to a plastic bag; close it perfectly in order to avoid any leaks.

This bag is protected of the light by covering it with a black bag. See Figure 1.

Start the ozone generator until the bag is filled (this takes around 3 min).

Sampling:

Simultaneously with filling the bag, prepare the reagent to detect the ozone: in a 10 mL Erlenmeyer flask pour an aliquot of 1 mL of KI solution and add 2 drops of acid and 2 drops of starch indicator. Suck this solution completely with the syringe.

Immediately after you have filled the bag, suck with the syringe 9 mL of the air in the bag. Shake

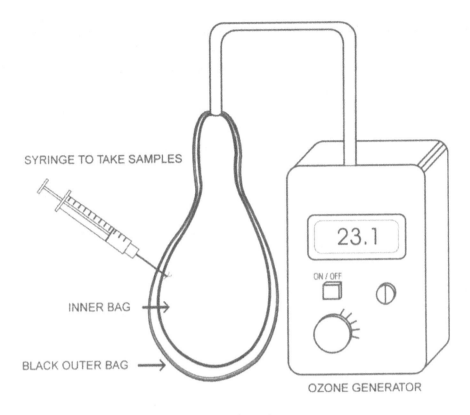

SYRINGE TO TAKE SAMPLES

23.1

ON / OFF

INNER BAG

BLACK OUTER BAG

OZONE GENERATOR

FIGURE 1. Experimental set-up.

the syringe with the mixture for 1 minute, and return it to the Erlenmeyer flask and titrate it with the thiosulfate solution. Register the reading of the used volume.

Take six or seven samples in total, at different time intervals (every 4 minutes or so) up to 20–25 minutes.

II) Ozone + CH_2Cl_2 vs time

Connect the exit of the ozone generator to a plastic bag; close it perfectly in order to avoid any leaks. This bag is protected of the light by covering it with a black bag.

Start the ozone generator until the bag is filled (this takes around 3 min). Once the bag is filled with air/ozone introduce 1 mL of methylene chloride with a syringe.

In a 10-mL Erlenmeyer flask pour a 1mL aliquot of KI solution and add 2 drops of acid and 2 drops of starch indicator. Suck this solution completely with the syringe.

Immediately after, take a 9-mL sample with a syringe containing 1 mL of the KI solution.

Shake the syringe with the mixture for 1 minute, return it to the Erlenmeyer flask and titrate it with the thiosulfate solution. Register the reading of the used volume.

Take six or seven samples in total, at different time intervals (every 4 minutes or so) up to 20–25 minutes.

III) Ozone + UV vs time

Repeat the previous procedure, but now place the UV lamp inside the bag before introducing the ozone. Start the ozone generator until the bag is filled (this takes around 3 min); once the bag is filled, turn on the UV lamp.

In a 10-mL Erlenmeyer flask pour a 1-mL aliquot of KI solution and add 2 drops of acid and 2 drops of starch indicator. Suck this solution completely with the syringe and then suck 9 mL of the air that is in the bag.

Shake the syringe with the mixture for 1 minute, transfer its contents to the 10-mL Erlenmeyer flask and titrate it with the thiosulfate solution.

Register the reading of the used volume. Take six or seven samples in total, at different time intervals (every 4 minutes or so) up to 20–25 minutes.

IV) Ozone + CH_2Cl_2 + UV vs time

Repeat the previous procedure for the filling of air/ozone in the bag. Once the bag is filled with air/ozone introduce 1 mL of methylene chloride with a syringe and turn on the UV lamp.

In a 10-mL Erlenmeyer flask pour a 1-mL aliquot of KI solution and add 2 drops of acid and 2 drops of starch indicator. Suck this solution completely with the syringe.

Immediately after, take a 9-mL sample with a syringe containing 1 mL of the KI solution.

Shake the syringe with the mixture for 1 minute, and return it to the Erlenmeyer flask. Then, titrate it with the thiosulfate solution. Register the reading of the used volume.

Take six or seven samples in total, at different time intervals (every 4 minutes or so) up to 20–25 minutes.

Name_____Section_____Date_____

Instructor_____Partner_____

PRELABORATORY REPORT SHEET—EXPERIMENT 10

Objectives

Flow sheet of procedure

Waste containment and recycling procedure

PRELABORATORY QUESTIONS AND PROBLEMS

1. Write the balanced equations for the reaction of ozone with KI.

2. Write the balanced equations for the reaction of iodine with thiosulfate.

3. Establish the proper equations to obtain the ozone concentration once you have the volume of titrant.

Additional Related Projects

- Perform the experiment with different quantities of methylene chloride.
- Perform the experiment with an olefin (i.e., ethylene) or acetylene.

Name_____Section_____Date_____

Instructor_____Partner_____

LABORATORY REPORT SHEET—EXPERIMENT 10

I) Ozone vs. Time

Time, min	Vol. titrant, mL	O$_3$ conc., mg/L
0		

III) Ozone + UV vs. Time

Time, min	Vol. titrant, mL	O$_3$ conc., mg/L
0		

II) Ozone + CH$_2$Cl$_2$ vs. Time

Time, min	Vol. titrant, mL	O$_3$ conc., mg/L
0		

IV) Ozone + CH$_2$Cl$_2$ + UV vs. Time

Time, min	Vol. titrant, mL	O$_3$ conc., mg/L
0		

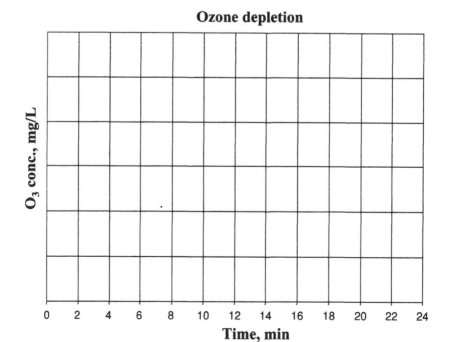

Ozone depletion

POSTLABORATORY PROBLEMS AND QUESTIONS

1. Discuss the results you obtained regarding how fast was the decreasing of ozone concentration under the various conditions.

2. Propose a different method for ozone quantification.

Student Comments and Suggestions

Literature References

Clesceri, L. S.; Greenberg, A.; Andrew E., Eds., *Standard Methods for the Examination of Water and Waste Water*, 20th ed.; American Public Health Association: Washington, D.C., 1998, Section 2, p. 42; Section 4, pp. 137–139.

Roan, S. *Ozone Crisis*; Wiley: New York, 1989.

Spiro, T. G.; Stigliani, W. M. *Chemistry of the Environment*, Prentice Hall: New Jersey, 1996.

Sponholtz, D. J.; Walters, M. A.; Tung, J.; BelBruno, J. J. "A Simple and Efficient Ozone Generator," *J. Chem. Educ.* **1999**, *76*, 1712–1713.

Experiment 11
Acid Mine (or Acid Rock) Drainage

Reference Chapters: 5, 8

Objectives

After performing this experiment, the student shall be able to:

- Understand key interactions between metal sulfides and their natural surroundings.
- Mimic mine tailings and observe their oxidation by different species.
- Test the role of Fe(III) as a natural oxidizer.

Introduction

Metal sulfides are ubiquitous in the environment. For example,

(a) They are the main components of ore rocks and are present in abandoned mining sites.
(b) Large coastal areas are covered with marine sulfide-rich sediments.
(c) Oceanic black smokers contain sizeable amounts of sulfides.

When sulfide minerals are exposed to air, they become oxidized. Microorganisms make this oxidation some 10^5–10^6 times faster. After coming in contact with water (mainly rainwater), these minerals form aqueous solutions that are notably acidic and are called *acid mine drainage* (*AMD*) or *acid rock drainage* (*ARD*). A wider classification of waters associated with mining projects is that of *mining-influenced waters* (*MIW*), as summarized in Table 1.

Because Fe is often the main metal present, AMD can be represented by the oxidation of pyrite:

$$FeS_{2(s)} + H_2O_{(1)} + 7/2\, O_{2(g)}$$
$$\rightarrow Fe^{2+} + 2SO_4^{2-} + 2H^+ \qquad (1)$$

The Fe^{2+} thus produced oxidizes to Fe^{3+} in air, albeit very slowly at low pH (with $t_{1/2}$ values on the order of years):

$$Fe^{2+} + {}^1\!/_4 O_{2(g)} + H^+ \rightarrow Fe^{3+} + {}^1\!/_2 H_2O_{(1)} \qquad (2)$$

Then, Fe^{3+} can either hydrolyze and form insoluble Fe(III) hydroxide, or act as an additional oxidant for FeS_2 to produce Fe^{2+}, as predicted from the corresponding Pourbaix diagram (see McNeil and Little, 1999):

$$FeS_{2(s)} + 14Fe^{3+} + 8H_2O_{(1)}$$
$$\rightarrow 15Fe^{2+} + 2SO_4^{2-} + 16H^+ \qquad (3)$$

The residues obtained after the main mineral or metal has been extracted are called *mine tailings*. They normally appear in a substantially divided form (as a result of crushing and grinding) and therefore have high surface areas that favor reaction 1 upon contact with air. The ensuing production of protons and sulfate ions yields sulfuric acid capable of dissolving minerals, thereby increasing metal and non-metal ion content and acidity in nearby surface waters, ground waters, and in the corresponding receiving water bodies. When the water table is near the surface, capillary rise and evaporation may also contaminate the upper soil layers. Amazingly, tailing dumps can be oxidized at depths of even several meters.

TABLE 1 Mining-Influenced Waters

Type	Abbrev.	pH	Primary treatment problem
Acid rock drainage (or acid mine drainage)	ARD (AMD)	Acidic	Elimination of acidity
Mineral processing water	MPW	Basic	Elimination of CN^-, AsO_4^{3-}, SeO_4^{2-}
Marginal waters	MW	Circumneutral	Removal of small conc. of contaminants in high flows of water
Residual waters	RW	Basic	Removal of high levels of TDS

Source: Wildeman and Schmiermund, 2004.

Schemes devised to ameliorate AMD production include the use of coatings to encapsulate the pyritic surface and the sequestration of Fe^{3+} with chelating agents, which reduces the effective concentration of this oxidant and can reduce its standard potential as well. Addition of sacrificial reducing agents (e.g., $CaSO_3$) may also prove helpful.

AMD processes typically occur in long time scales, and as such this subject is hardly fit for school laboratory experiments. The handling of the necessary bacteria would introduce additional difficulties. Nonetheless, because these phenomena are important from the perspective of environmental education, we present here suitable conditions to simulate and study the (non-biological) production of AMD in a laboratory session using simple equipment and materials.

Experimental Procedure

Estimated time to complete the experiment: 4 h. (Note: In order to complete the five experiments suggested below in a single lab session, we recommend that five similar setups be prepared and run simultaneously by different student teams).

Materials	Reagents
1 pH meter	nitrogen gas
4 20-mL beakers	FeS (solid)
1 magnetic stirrer	D. I. water
1 aquarium pump	ferric sulfate
1 2-mL graduated pipet	
1 spatula	
2 Beral pipets	

Note: The term *pyrite* commonly refers to the mineral FeS_2. Ferrous sulfide, FeS is a precursor

of FeS_2 formation, resulting from the interaction of iron minerals with the products of biological sulfate reduction under anoxic conditions. Because the rate of oxidation of the latter may be greater than that of pyrite and it is commercially accessible, we selected FeS for the following experiments.

Experiment 1

Set up a system consisting of a 20-mL beaker equipped with a pH electrode, a small magnetic stirring bar, and a small-diameter plastic or rubber tube equipped with a Pasteur pipet or a glass tube (so as to bubble either air or nitrogen to a reaction mixture contained in the beaker). See Figure 1. Grind a few small pieces of commercial FeS with a mortar and pestle. Because grinding produces different degrees of surface oxidation, the instructor may wish to establish a "grinding protocol." **Caution: Even though FeS is not generally regarded as a health hazard, it is advisable not to breathe its dust.**

Add 100 mg of the resulting powder to 10 mL of deionized water inside the beaker. Now, begin bubbling air by means of an aquarium pump (we have used a flow rate of 80 mL min^{-1}) and turn on the stirrer. After a short equilibration period (e.g., 10 min, during which the pH is somewhat erratic), start monitoring pH as a function of time for at least 120 minutes (or more, if time permits). Plot the resulting pH vs time curve.

Experiment 2

In order to simulate reaction 3, repeat *Experiment 1* but this time also add 10 mg of ferric sulfate to the reaction mixture and then monitor the pH as above.

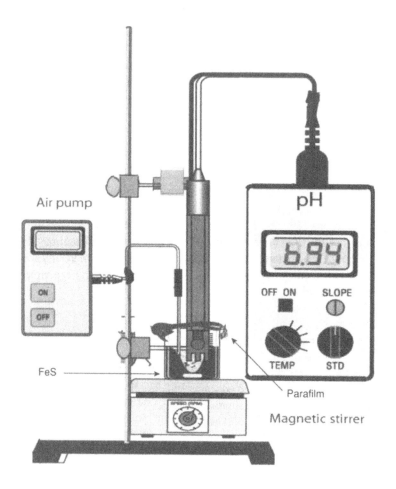

FIGURE 1. Experimental set-up.

Experiment 3

Repeat *Experiment 2*, but do not include FeS in the reaction mixture.

Experiment 4

Repeat *Experiment 2*, but use nitrogen gas instead of air.

Experiment 5

Repeat *Experiment 4*, but use 1 mL of fresh 30% H_2O_2 instead of $Fe_2(SO_4)_3$.

Caution: 30% H_2O_2 is corrosive and a strong oxidizer. Use protection to avoid contact with human tissue.

Name_____Section_____Date_____

Instructor_____Partner_____

PRELABORATORY REPORT SHEET—EXPERIMENT 11

Objectives

Flow sheet of procedure

Waste containment and recycling procedure

PRELABORATORY QUESTIONS AND PROBLEMS

*1. The reaction responsible for the production of acidity during the acid mine drainage process of oxidation of pyrite by dioxygen can also be written as:

$$aFeS_{2(s)} + bO_{2(g)} + cH_2O_{(l)}$$
$$\rightarrow aFe(OH)_{3(s)} + dSO_4^{2-} + eH^+$$

Find the smallest integer values for a, b, c, d, and e so as to balance this equation.

*2. Calculate the *acid-producing potential* of pyrite, defined as the number of millimoles of protons theoretically produced by 100 g of a pure pyrite sample (see Horan, 2005), based on the overall reaction found in Problem 1.

_____

* Answer in this book's webpage at www.springer.com

*3. Calculate the pH of a 0.1 M solution of $Fe_2(SO_4)_3$.

*4. Calcium carbonate is sometimes used for neutralizing acid mine drainage. Calculate the *neutralizing capacity* of 100 g of $CaCO_3$, in terms of the millimoles of H^+ that this amount can theoretically neutralize (see Horan, 2005).

5. Explain the difference between mineral acidity and proton acidity in an AMD.

Additional Related Projects

- Titrate the solutions resulting from each experiment and calculate the moles of H^+ produced. Because acid mine drainage is sometimes neutralized with $CaCO_{3(s)}$, calculate the number of moles of this substance needed to neutralize each solution.

- Monitor the concentrations of Fe(II) and Fe(III) with time and compare them with the concentration of protons (see Kargbo, 2004).

- Use microwaves to speed-up the dissolution process and compare to a control without microwaves (see Kuslu, 2002).

- For a quick demonstration of the acidifying effect of metal sulfide oxidation, test the pH of a drop of 30% H_2O_2 using pH paper. Place a small amount of FeS_2 or FeS (about 10 mg) in a small test tube and add 15 drops of 30% H_2O_2 (see the cautionary note about H_2O_2 above). Allow the reaction to proceed until the bubbling stops. Test the pH of the solution. (See Horan, 2005).

- Because it is known that the oxygen in the sulfates resulting from acid mine drainage comes from the water in contact with pyrite, try a non-oxygenated, polar solvent and observe and comment on any possible differences in behavior observed (especially rates), as compared to your experimental observations from the present lab experiment. (See Usher, 2004).

- Treat a simulated acid mine waste by preparing a solution containing Fe^{3+}, Fe^{2+}, and Zn^{2+} (400 mg, 400 mg, and 130 mg per liter, respectively), neutralizing it with NaOH, and precipitating the metal sulfides with Na_2S. Analyze the metal concentrations in the solution before and after the treatment. (See Horan, 1997 and 2005).

- Acid mine waste is sometimes treated with different organic wastes—used as substrates—whereby biological, chemical, and physical processes aid in its clean up. Use different wastes (e.g., compost, manure, hay, rich soil, and grass) and place them in a solution to be treated (prepared as in the previous project), both in an open and in a closed container. This will allow aerobic and anaerobic processes to be compared. Observe, analyze, and compare both samples. Comment on your findings. (See Horan, 1997 and 2005).

Name_____Section_____Date_____

Instructor_____Partner_____

LABORATORY REPORT SHEET—EXPERIMENT 11

Source and purity of the FeS used   _____

Experiment 1
Volume of solution   _____mL
Weight of FeS   _____mg
Sieve size used with FeS (optional)   _____
Carrier gas   _____
Flow rate of the carrier gas   _____mL min^{-1}
Plot your pH vs time data.

t, min

pH

Experiment 2
Moles of ferric sulfate added   _____moles
Carrier gas   _____
Flow rate of the carrier gas   _____mL min^{-1}
Plot your pH vs time data.

pH

t, min

Experiment 3
Carrier gas   _____
Flow rate of the carrier gas   _____mL min^{-1}
Plot your pH vs time data.

pH

t, min

Experiment 4
Carrier gas   _____
Flow rate of the carrier gas   _____mL min^{-1}
Plot your pH vs time data.

pH

t, min

Experiment 5
Volume and concentration of the H$_2$O$_2$ _____mL, _____%
Carrier gas   _____
Flow rate of the carrier gas   _____mL min^{-1}
Plot your pH vs time data.

pH

t, min

Compare the five curves obtained on the same plot. Are the results what you expected?

Explain the behavior of each curve.

POSTLABORATORY PROBLEMS AND QUESTIONS

*1. During pyrite oxidation at low pH, in the presence of Fe^{3+} but in the absence of O_2, the system obeys the relationship

$$\log (M/M_0) = -kt$$

where M_0 and M are the number of surface moles of unreacted FeS_2 at the beginning of the experiment and at a time t, respectively, and k is a constant. (See Kargbo, 2004).

a) Obtain from this expression the rate equation for pyrite oxidation under these conditions in the form of dM/dt as a function of M.
b) What is the reaction order?

*2. Pyrite oxidation is a very complex process during which its surface becomes covered with different insoluble species. These can be analytically discriminated and the following table gives the (rounded off) elemental ratios thus found or estimated. The species involved are FeO, Fe_2O_3, Fe_3O_4, FeOOH, $Fe(OH)_2$, $Fe(OH)_3$, and FeS_2. Assign the formula of each species to its corresponding data. (Note that sulfur in pyrite bears an oxidation state of -1).

In the example solved in the table for Fe_3O_4, one can note that this species results from the combination of one FeO unit and one Fe_2O_3 unit. The ratio of Fe(III) to oxygen-containing ions, ($OH^- + O^{2-}$) is then $2/4 = 0.50$ (because there are no OH^- in Fe_3O_4). The total number or irons is 3 and the total number of oxide ions is 4. Then, the Fe(Tot)/($OH^- + O^{2-}$) ratio is $3/4 = 0.75$.

Fe(III)/ $(OH^- + O^{2-})$	Fe(Tot)/ $(OH^- + O^{2-})$	$O^{2-}/$ OH^-	Fe(II)/ $S(-I)$	Species
0.50	0.75	—	—	Fe_3O_4
—	0.50	—	—	
—	—	—	0.50	
—	1.00	—	—	
0.67	0.67	—	—	
0.50	0.50	1.00	—	
0.33	0.33	—	—	

Data obtained by analyzing the sizes and location of the peaks corresponding to the electron binding energies of their constituent atoms with XPS (X-ray photoelectron spectroscopy) (Bonnissel-Gissinger, 1998).

*3. A long-standing question concerning the mechanism of pyrite oxidation in the presence of dioxygen in wet environments is whether the oxygen that becomes incorporated into the resulting SO_4^{2-} and FeOOH species (see the corresponding reaction in the Introduction) comes from the dissolved dioxygen or from the water.

Using an a analytical technique (sensitive to isotopic-substitution) with $H_2^{16}O$, $H_2^{18}O$, and $^{16}O_2$, control tests were run with pure reagents prepared with and without isotope-labeled water and dioxygen. The results were then compared to those obtained with pyrite samples also exposed to isotope-labeled water and to dioxygen. The results are given in the table. (Note that $^{16}O_2$ was used in all cases).

_____

* Answer in this book's webpage at www.springer.com

Group analyzed	Control tests				Sample tests			
	SO_4^{2-}		FeOOH		SO_4^{2-}		FeOOH	
Water used	$H_2^{16}O$	$H_2^{18}O$	$H_2^{16}O$	$H_2^{18}O$	$H_2^{16}O$	$H_2^{18}O$	$H_2^{16}O$	$H_2^{18}O$
Peak observed, cm^{-1}	1105	1030	835	below 800	1105	1030	835	835

Data obtained with *in situ* reflectance infrared spectroscopy (Usher, 2004).

Complete the following phrase by selecting the option from the list below that best interprets these results:

"The oxygen that becomes incorporated into SO_4^{2-} comes from _____, and the oxygen that becomes incorporated into FeOOH comes from_____".

Options:

  i) water, water

  ii) dioxygen, dioxygen

  iii) water, dioxygen

  iv) dioxygen, water

Student Comments and Suggestions

Literature References

August, E. E.; McNight, D. M.; Hrncir, D. C.; Garhart, K. S. "Seasonal Variability of Metals Transport Through a Wetland Impacted by Mine Drainage in the Rocky Mountains," *Environ. Sci. Technol.* **2002**, *36*, 3779–3786.

Batty, L. C.; Younger, P. L. "Critical Role of Macrophytes in Achieving Low Iron Concentrations in Mine Water Treatment Wetlands," *Environ. Sci. Technol.* **2002**, *36*, 3997–4002.

Bonnissel-Gissinger, P.; Alnot, M.; Ehrhardt, J.-J.; Behra, P. "Surface Oxidation of Pyrite as a Function of pH," *Environ. Sci. Technol.* **1998**, *32*, 2839–2845.

Chernyshova, I. V. "An in Situ FTIR Study of Galena and Pyrite Oxidation in Aqueous Solution," *J. Electroanal. Chem.* **2003**, *558*, 83–98.

Hansen, J. P.; Jensen, L. S.; Bedel, S.; Dam-Johansen, K. "Decomposition and Oxidation of Pyrite in a Fixed-Bed Reactor," *Ind. Eng. Chem. Res.* **2003**, *42*, 4290–4295.

Hao, Y.; Dick, W. A. "Potential Inhibition of Acid Formation in Pyritic Environments Using Calcium Sulfite Byproduct," *Environ. Sci. Technol.* **2000**, *34*, 2288–2292.

Horan, J. H.; Wildeman, T. R. Dept. of Chemistry and Geochemistry, Colorado School of Mines. "Environmental Chemistry in Colorado Toxic Mine Drainage: Chemistry and Treatment," 2005, http://www.mines.edu/fs_home/jhoran/ch126/index.htm

Horan, J. H.; Wildeman, T. R.; Ernst, R. "Acid Mine Drainage Laboratory Experiments," 214th. Am. Chem. Soc. Nat. Meet., Chem. Ed. Division Paper # 005. Las Vegas, NV, Sept. 7–11, 1997.

Howard, A. G. *Aquatic Environmental Chemistry*; Oxford Chemistry Primers. Oxford Science Publications: Oxford, 1998. Chapter 5.

Ibanez, J. G. "Microscale Environmental Chemistry. Part 6. Water Acidification by Oxidation of Mineral Sulfides (Acid Mine Drainage)," *Chem. Educ.* **2006**, *11*, 251–253.

Kargbo, D. M.; Atallah, G.; Chatterjee, S. "Inhibition of Pyrite Oxidation by a Phospholipid in the Presence of Silicate," *Environ. Sci. Technol.* **2004**, *38*, 3432–3441.

Kuslu, S.; Bayramoglu, M. "Microwave-Assisted Dissolution of Pyrite in Acidic Ferric Sulfate Solution," *Ind. Eng. Chem. Res.* **2002**, *41*, 5145–5150.

McNeil, M. B.; Little, B. J. "The Use of Mineralogical Data in Interpretation of Long-Term Microbiological Corrosion Processes: Sulfiding Reactions," *J. Am. Inst. Conserv.* **1999**, *38*, 186–199.

Naicker, K.; Cukrowska, E.; McCarthy, T. S. "Acid Mine Drainage Arising from Gold Mining Activity in Johannesburg, South Africa and Environs," *Environ. Poll.* **2003**, *122*, 29–40.

News of the Week. *Chem. & Eng. News.* Nov. 10, 2003, p. 16.

Nordstrom, D. K.; Alpers, C. N.; Ptacek, C. J.; Blowers, D. W. "Negative pH and Extremely Acidic Mine Waters from Iron Mountain, California," *Environ. Sci. Technol.* **2000**, *34*, 254–258.

Rawls, R. "Some Like it Hot", *Chem. Eng. News* Dec. 21, 1998. p. 35–39.

Singer P. C.; Stumm, W. "Acidic Mine Drainage: The Rate Determining Step," *Science* **1970**, *167*, 1121–1123.

Sundstrom, R.; Astrom, M.; Osterholm, P. "Comparison of the Metal Content in Acid Sulfate Soil Runoff and Industrial Effluents in Finland," *Environ. Sci. Technol.* **2002**, *36*, 4269–4272.

Usher, C. R.; Cleveland, C. A. Jr.; Strongin, D. R.; Schoonen, M. A. "Origin of Oxygen in Sulfate during Pyrite Oxidation with Water and Dissolved Oxygen: An In Situ Horizontal Attenuated Total Reflectance Infrared Spectroscopy Isotope Study," *Environ. Sci. Technol.* **2004**, *38*, 5604–5606.

Wildeman, T. R.; Schmiermund, R. "Mining-Influenced Waters: Their Chemistry and Methods of Treatment," 2004 National Meeting of the American Society of Mining and Reclamation and The 25th West Virginia Surface Mine Drainage Task Force, April 18–24, 2004.

Experiment 12
Electrochemical Treatment of Gas Pollutants

Reference Chapter: 10

Objectives

After performing this experiment, the student shall be able to:

- Convert a gas pollutant into its constituent elements in a pure and useful form.
- Understand and apply an electrochemical indirect method for the destruction of a (hazardous) waste gas.
- Interpret the physical and chemical phenomena (i.e., color changes, bubbling) observed during the course of an indirect electrochemical reaction.

Introduction

Electrochemical remediation methods require an ion-conducting medium to perform their function of oxidizing or reducing polluting species. For this reason, gaseous mixtures must be normally absorbed first in aqueous solutions to be treated. This can be accomplished either by using an absorption medium inside an electrochemical cell (*inner-cell process*) or by first absorbing the gas and then transferring the absorption medium into the electrochemical cell for treatment (*outer-cell process*). Also, in such methods the polluting species can either undergo electron transfer on an electrode surface (*direct electrolysis*), or electrons can be shuttled to/from the electrode by an electron carrier or *mediator* (*indirect electrolysis*). Some examples of gases treated electrochemically are discussed in Chapter 10.

Hydrogen sulfide is a well-known pollutant produced in considerable amounts by sulfate reduction in organic-rich (anaerobic) environments, from heavy oil desulfurization processes, oil recovery operations, coal gasification/liquefaction processes, etc. It is frequently treated by absorption in basic scrubbing solutions of different amines whereby the scrubbing liquor can be regenerated, providing a concentrated stream of H_2S that unfortunately requires further treatment. Alternatively, the Claus process can be used:

$$H_2S_{(g)} + \tfrac{3}{2}O_{2(g)} \rightarrow SO_{2(g)} + H_2O_{(g)} \qquad (1)$$

$$H_2S_{(g)} + SO_{2(g)} \rightarrow 2S_{(s)} + H_2O_{(g)} + \tfrac{1}{2}O_{2(g)} \qquad (2)$$

This presents the following disadvantages: 1) hydrogen is essentially wasted, because H_2O is produced from it, 2) the high temperatures and catalysts required do not offer flexible adjustment to varying concentrations of H_2S, 3) a pre-treatment for the separation of companion hydrocarbons and H_2 is required, and 4) a post-treatment is required since the Claus process converts only 90–98% of the initial H_2S content.

In this experiment, H_2S will be produced and then absorbed with simultaneous reaction, whereby sulfide ions are oxidized by a chemical oxidant (mediator) to elemental sulfur. Then, elemental hydrogen will be produced at the cathode of an electrolytic cell and the oxidant will be simultaneously regenerated at the anode. The net result is rather uncommon: the decomposition of a pollutant (H_2S) into its components in their pure, useful elemental forms.

The reactions are:

a) Chemical:

$$H_2S_{(g)} + I_3^- \rightarrow S_{(s)} + 3I^- + 2H^+$$
$$\text{(in the aqueous solution)} \qquad (3)$$

b) Electrochemical:

$$3I^- \rightarrow I_3^- + 2e^- \quad \text{(at the anode)} \quad (4)$$

$$2H^+ + 2e^- \rightarrow H_{2(g)} \quad \text{(at the cathode)} \quad (5)$$

Experimental Procedure

Estimated time to complete the experiment: 1.5 h

Materials	Reagents
2 20-mL beakers	H$_2$S
1 funnel	Solution 0.25 M KI, 0.05 M
1 2-mL graduated pipet	I$_2$, 0.05 M HCl
1 5-mL graduated pipet	6 M HCl
2 Beral pipets	ZnS
filter paper	D. I. water
1 platinum electrode	
1 lead electrode	
1 graphite electrode	
2 alligator clips	
1 9-V battery, or an AC/DC	
converter	

This experiment is comprised of two steps. (1) A chemical oxidation step, where H$_2$S (or a sulfide in solution) is oxidized by I$_2$/I$_3^-$, and (2) the electrochemical regeneration of the I$_2$/I$_3^-$ by oxidation of the I$^-$ with the concomitant reduction of H$^+$ to H$_2$. See Figure 1.

Prepare with stirring a stock solution to be 0.25 M in KI, 0.05 M in I$_2$, and 0.05 M in HCl. This will be called the *scrubbing solution*. Pour 2 mL of this solution into a 3-mL conical bottom vial. Under a well-ventilated hood bubble H$_2$S gas through it, produced by conventional methods (for example, by adding 2–3 mL of 6 M HCl to 0.11 g of ZnS in a 30- or 60-mL syringe, according to the method of Mattson, 2001). **Caution: The acid is corrosive to human tissue, handle with care.** Alternatively, add slowly 5 to 7 drops of a concentrated sulfide solution (ammonium sulfide works well) that simulates the presence of H$_2$S in the scrubbing solution.

Caution: H$_2$S is a very poisonous gas. It binds to hemoglobin where dioxygen should be bound, and thus prevents its uptake and transport. It can lead to intoxication (and in extreme cases, even to death). Prepare and use it under the hood. Sulfide solutions must be handled with the same precautions.

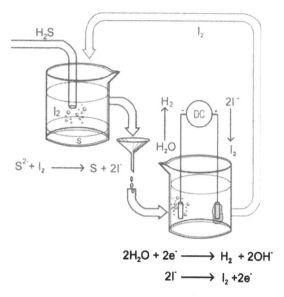

$$S^{2-} + I_2 \longrightarrow S + 2I^-$$

$$2H_2O + 2e^- \longrightarrow H_2 + 2OH^-$$

$$2I^- \longrightarrow I_2 + 2e^-$$

FIGURE 1. Indirect electrochemical treatment of H$_2$S. (Adapted from Ibanez, 2006).

A dramatic change in color will occur within a few minutes due to the oxidation of S^{2-} ions by I$_2$ (or better, by the I$_3^-$ ions), producing a suspension of yellow elemental sulfur plus colorless iodide ions. Filter this suspension with a very fine filter paper and collect the filtrate in a 10-mL beaker. This will serve as the electrochemical cell for this experiment. The resulting filtrate will now be electrolyzed with a 9 V battery or an electronic direct power source. To this end, partially dip two electrodes in the solution (for example, platinum, lead, or graphite). Do not let them touch each other! Connect each one with an alligator clip to the terminals of the power source.

Turn the power on. If a regulated power source is used, maintain a voltage between 1 and 3 volts. **Note:** If graphite is used, maintain the voltage as low as possible (just enough as to see a color change, as described below), because otherwise the graphite at the anode may disintegrate in pieces due to the production of O$_2$ which diffuses into its basal planes, causing their rupture. Also, if a chlorine-like smell is noticed, decrease the applied potential since chloride ions from the HCl present in the solution can be oxidized to produce chlorine gas.

After a short time, a dramatic color change is noted in the vicinity of the anode (i.e., it goes back to dark brown) as a result of the re-oxidation of the I^- ions back to I_2 or I_3^-. High purity hydrogen is simultaneously produced at the cathode, which can then be collected and tested in the traditional manner.

The iodine solution is now ready to be re-utilized for sulfide oxidation if desired, which makes this experiment even more appealing from an environmental perspective. (For example, we have performed at least 20 of these cycles with the same solution).

Name _____ Section _____ Date _____

Instructor _____ Partner _____

PRELABORATORY REPORT SHEET—EXPERIMENT 12

Objectives

Flow sheet of procedure

Waste containment and recycling procedure

PRELABORATORY QUESTIONS AND PROBLEMS

1. Give three examples of cases where electrochemical techniques can, in principle, be utilized for the transformation and/or removal of pollutants.

2. If an electrochemically-active pollutant is to be oxidized and it has a standard potential of $+0.50$ V (vs the standard hydrogen electrode, SHE), which of the following potentials could, in principle, be applied to an electrode to efficiently achieve that operation?

a) -0.50 V, b) 0 V, c) $+0.50$ V, d) $+1.00$ V vs SHE. Explain.

*3. Formaldehyde is a well-known toxic organic substance frequently found as a vapor. It can be

degraded into harmless products with electrogenerated hydrogen peroxide either in acidic (with the aid of a catalyst, not shown below) or basic solutions according to the following equations, respectively:

a) $O_2 + 2H^+ + 2e^- = H_2O_2$

$HCHO + H_2O_2 = HCOOH + H_2O$

b) $O_2 + H_2O + 2e^- = HO_2^- + OH^-$

$2HCHO + HO_2^- + OH^-$
$= 2HCOO^- + H_2O + H_2$

Based only on stoichiometric arguments and assuming no side reactions, in which of the two cases is the current efficiency for formaldehyde destruction higher? (Note: Current efficiency can be defined as the charge employed for a given process divided by the total charge passed through the system).

_____

* Answer in this book's webpage at www.springer.com

Additional Related Projects

- Replace the I_2/I^- oxidizing system with Fe^{3+}/Fe^{2+}.
- Perform the reactive absorption of SO_2 by bubbling it through a solution containing dissolved Fe^{3+}. The H_2SO_4 thus produced can be analyzed (e.g., quantitatively by titrimetry, qualitatively by any standard test for sulfates). (See Rajeshwar, 1997).
- Absorb SO_2 in an electrochemical cell containing solid Cu particles in a conductive solution. Dioxygen dissolved in the aqueous phase can oxidize SO_2 to H_2SO_4 in the presence of Cu, which in turn produces Cu^{2+} ions that are reduced at the cathode. Anodic oxidation of SO_2 can also occur, which increases the conversion efficiency. (See Rajeshwar, 1997).

Name _____ Section _____ Date _____

Instructor _____ Partner _____

LABORATORY REPORT SHEET—EXPERIMENT 12

Observations

1. Original color of the solution   _____

2. Moles of sulfide used in the syringe   _____ moles

3. Theoretical number of moles of gas produced   _____ moles

4. Moles of iodine used   _____ moles

5. Color of the aqueous mixture after bubbling the H_2S gas _____

6. Color observed at the *anode* during electrolysis   _____

7. Color observed at the *cathode* during electrolysis   _____

POSTLABORATORY PROBLEMS AND QUESTIONS

1. Find in the literature an example of a *direct* electrochemical treatment of a polluting gas.

*2. The general process for the electrocatalytic oxidation of an organic molecule $(C_nH_xO_y)$ by Ag(II) ions in the form of $AgNO_3^+$ is as follows. First, Ag(I) is oxidized to Ag(II) at the anode of an electrochemical cell. Then, Ag(II) oxidizes the organic to CO_2:

$$aAgNO_3^+ + C_nH_xO_y + bH_2O$$
$$= aAg^+ + nCO_2 + aHNO_3$$

and the cycle can be repeated again. For example, for the oxidation of the two carbons in acetic acid to CO_2 (where the carbon atom is in an oxidation state of +4), eight electrons need to be removed, whereas the same procedure for ethanol requires removal of 12 e^-.

With this information write the above equation for the following substances and balance the resulting equations: ethylene glycol, acetone, and benzene.

*3. The electrocatalytic reduction of CO_2 to produce n-alcohols, R-OH can occur as follows:

$$aCO_2 + bH_2O + ne^- = R\text{-}OH + cOH^-$$

where R is a C_1, C_2, or C_3 aliphatic carbon chain.
a) Write the balanced equations for the production of methanol, ethanol, and *n*-propanol.
b) Find three generalized algebraic equations that relate the stoichiometric coefficients as follows: n with c, n with a, and n with b. [In other words, $n = f(c)$, $n = f'(a)$, $n = f''(b)$].

_____

* Answer in this book's webpage at www.springer.com

Student Comments and Suggestions

Literature References

Bockris, J. O'M.; Reddy, K. *Modern Electrochemistry, Vol. 2*; Plenum Press: New York, 2000. Chapter 15.

Genders, J. D.; Weinberg, N., Eds. *Electrochemistry for a Cleaner Environment*; The Electrosynthesis Co.: E. Amherst, NY, 1992.

Glasscock, D. A.; Rochelle, G. T. "Approximate Simulation of CO_2 and H_2S Absorption into Aqueous Alkanolamines," *AIChE Journal* **1993**, *39*, 1389–1397.

Ibanez, J. G. "Laboratory Experiments on the Electrochemical Remediation of the Environment. Part 5: Indirect H_2S Remediation," *J. Chem. Educ.* **2001**, *78*, 778–779.

Ibanez, J. G. "Electrochemistry for Environmental Remediation: Laboratory Experiments," *Educ. Quim.* (Mexico, in English), **2006**, *17*, 274–278.

Ibanez, J. G.; Takimoto, M.; Vasquez, R.; Rajeshwar, K.; Basak, S. "Laboratory Experiments on Electrochemical Remediation of the Environment: Electrocoagulation of Oily Wastewater," *J. Chem. Educ.* **1995**, *72*, 1050–1052.

Ibanez, J. G.; Singh, M. M.; Szafran, Z.; Pike, R. M. "Laboratory Experiments on Electrochemical Remediation of the Environment. Part 2: Microscale Indirect Electrolytic Destruction of Organic Wastes," *J. Chem. Educ.* **1997**, *74*, 1449–1450.

Ibanez, J. G.; Singh, M. M.; Szafran, Z.; Pike, R. M. "Laboratory Experiments on Electrochemical Remediation of the Environment. Part 3. Microscale Electrokinetic Processing of Soils," *J. Chem. Educ.* **1998**, *75*, 634–635.

Ibanez, J. G.; Singh, M. M.; Szafran, Z.; Pike, R. M. "Laboratory Experiments on Electrochemical Remediation of the Environment. Part 4. Color Removal of Simulated Wastewater by Electrocoagulation-Electroflotation," *J. Chem. Educ.* **1998**, *75*, 1040–1041.

Kalina, D. W.; Maas, Jr., E. T. "Indirect Hydrogen Sulfide Conversion. 1. An Acidic Electrochemical Process," *Int. J. Hydrog. Energy* **1985**, *10*, 157–162.

Kalina, D. W.; Maas, Jr., E. T. "Indirect Hydrogen Sulfide Conversion. 2. A Basic Electrochemical Process," *Int. J. Hydrog. Energy* **1985**, *10*, 163–167.

Mattson, B. *Microscale Gas Chemistry*, 2nd. ed.; Educational Innovations: Norwalk, CT, 2001. p. 92.

Mizuta, S.; K., W.; Fujii, K.; Iida, H.; Isshiki, S.; Noguchi, H.; Kikuchi, T.; Sue, H.; Sakai, K. "Hydrogen Production From Hydrogen Sulfide By The Fe-Cl Hybrid Process," *Ind. Eng. Chem. Res.* **1991**, *30*, 1601–1608.

Rajeshwar, K.; Ibanez, J. G. *Environmental Electrochemistry: Fundamentals and Applications in Pollution Abatement*; Academic Press: San Diego, CA, 1997. Chapter 5.

Rajeshwar, K.; Ibanez, J. G.; Swain, G. M. "Electrochemistry and the Environment," *J. Appl. Electrochem.* **1994**, *24*, 1077–1091.

Sequeira, C. A. C., Ed. *Environmentally Oriented Electrochemistry*; Studies in Environmental Science #59; Elsevier: Amsterdam, 1994.

Smet, E.; Lens, P.; Van Langenhove, H. "Treatment of Waste Gases Contaminated with Odorous Sulfur Compounds," *Crit. Rev. Environ. Sci. Technol.* **1998**, *28*, 89–117.

Tarver, G. A.; Dasgupta, P. K. "Oil Field Hydrogen Sulfide in Texas: Emission Estimates and Fate," *Environ. Sci. Technol.* **1997**, *31*, 3669–3676.

Weinberg, N. L.; Genders, J. D.; Minklei, A. O. "Methods for Purification of Air," U.S. Patent 5,009,869 (April 23, 1991).

Experiment 13
Electrochemical Treatment of Liquid Wastes

Reference chapter: 10

Objectives

After performing this experiment, the student shall be able to:

- Understand and apply an indirect oxidation method for the destruction of hazardous wastes.
- Interpret the changes observed during the course of an indirect electrochemical reaction.
- Construct an electrochemical cell.

Introduction

As discussed in Chapter 10, electrochemical processes can often be used to address environmental problems. Some of the main characteristics that make these techniques promising include their versatility, environmental compatibility, economic feasibility, ambient temperature and pressure requirements, and amenability to automation. Electrochemical approaches to remediation include direct and indirect electrolysis, membrane-based approaches (e.g., electrodialysis), electrokinetic remediation of soils, dissolution-electrolysis of gases, photoelectrochemical techniques, disinfection of water, etc.

The electrolytic production of an intermediate species for the oxidation/reduction of another substance is known as *indirect electrolysis* (this concept is discussed in Section 10.1). This process can be made reversible or irreversible, depending on whether the mediator (catalyst) can or cannot be electrochemically regenerated and recycled in the process. The equations that describe the reversible

case (for an oxidation reaction) are:

$$C \rightarrow C^+ + e^- \qquad E_1^0 \quad (1)$$
$$C^+ + \text{Red} \rightarrow C + \text{Ox}^+ \qquad E_2^0 \quad (2)$$

where C is the redox mediator, Red is the pollutant species, Ox^+ is its oxidized product and E_1^0 and E_2^0 are the corresponding standard potentials for reactions 1 and 2 (here, $E_2^0 > 0$). In addition to the reduction or oxidation of the target pollutant, other reactions can compete. The most likely cathodic parasitic reactions (reductions) in the presence of dioxygen in aqueous solutions are:

(acidic solution): $2H^+ + 2e^- = H_2$
$$E^0 = 0.00 \text{ V} \quad (3a)$$
(neutral or basic): $\frac{1}{2} O_2 + H_2O + 2e^- = 2OH^-$
$$E^0 = 0.401 \text{ V} \quad (3b)$$

whereas the most common competing anodic reaction (oxidation) is the reverse of:

$$O_2 + 4H^+ + 4e^- = 2H_2O \qquad E^0 = 1.23 \text{ V} \quad (4)$$

In this experiment, students will produce a strong oxidizer [i.e., Co(III)] at the anode of a home-made electrolytic cell. This species will, in turn, oxidize a surrogate organic pollutant (e.g., ethylene glycol, ethanol, glycerol) to produce CO_2. The resulting gas is collected in a trap and its presence made evident by the appearance of a precipitate. Once reduced by its oxidizing action upon the pollutant, the mediator can be electrochemically regenerated so as to initiate the treatment cycle.

Experimental Procedure

Estimated time to complete the experiment: 1.5 h

Materials	Reagents
1 1-mL graduated pipet	$Ca(OH)_2$
3 Beral pipets	$CoSO_4 \cdot 7H_2O$
1 collection tube	H_2SO_4
1 10 × 75 mm test tube	D. I. water
2 nichrome wires or paper clips	glycerin
1 Pasteur pipet	acetic acid (or vinegar)
1 10-mL beaker	dilute H_2SO_4
1 50-mL beaker	$CaCO_3$
1 water bath	
1 Cu wire	
Hot silicon glue	
1 9-V battery, or an AC/DC converter	
1 thermometer	
2 alligator clips	
1 rubber stopper	

Experimental steps: The procedure involves three steps: (1) oxidation of a surrogate organic waste (glycerin) with a redox mediator in aqueous solution, (2) scrubbing of the gases exiting from the electrochemical cell to trap the CO_2, and (3) regeneration of the reduced half of the redox mediator, Co(II) by addition of an oxidizable organic compound to the remaining Co(III) solution.

Obtain a disposable plastic transfer pipet (also called Beral pipet) of approximately 4–5 mL capacity to be used as a microelectrochemical cell. Cut 6–7 cm (approx. 2.5 in) of Pb wire (alternatively, a small piece of thin Pb foil can be "rolled" as to make a wire replacement) to act as the anode and a similar length of either nichrome wire or a paper clip (normally made of steel) to serve as the cathode. Pt wire can also be used as the anode. Insert both electrodes in opposite sides of the upper portion of the Beral pipet in such a way that they do not touch each other as shown in Figure 1. Prepare a clear saturated solution of $Ca(OH)_2$ by stirring approximately 0.1 g of $Ca(OH)_2$ in 10 mL of deionized water followed by filtration.

Note: Avoid any danger posed by Pb residues possibly left on your hands, by washing thoroughly with soap and water.

Obtain from the instructor or prepare a solution of 0.05 M $CoSO_4$ in 2 M H_2SO_4. This can be prepared by adding 70 mg of $CoSO_4 \cdot 7H_2O$ to 5 mL of 2 M H_2SO_4 in a 10-mL beaker (stir well for a few minutes until all the solid is dissolved). Place 2–3 mL of this solution in another 10-mL beaker. Prepare in a 10 × 75 mm test tube a solution of a non-volatile surrogate organic waste by adding 1 drop of glycerin to 50 drops of water. Mix and take one drop of this solution and add it to the cobalt–sulfuric acid solution (this contains approximately 1.4×10^{-2} mmoles or 1 μL of glycerin). Alternatively, acetic acid (or vinegar) can be used. In this case, add, for example, four drops of a 6 M acetic acid solution directly to the cobalt–sulfuric acid solution. Shake for a few seconds as to have the surrogate waste completely dissolved. Then, either draw this pink solution into the Beral pipet through its stem or introduce it with a disposable Pasteur pipet or a small syringe through one of the holes punctured by the wires (then, reinsert the removed wire). To prevent any gas leaks (see below), a silicon-type sealant or melted wax may be applied at the electrode insertion points. Next, place the saturated $Ca(OH)_2$ solution in a collection tube (e.g., a small culture tube). Insert the tip of the stem of the pipet into this solution so that the gases exiting from the end of the stem of the Beral pipet bubble through it. See Figure 1. In order to enable the bending of the pipet stem and its

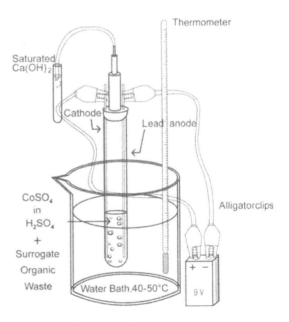

FIGURE 1. Microelectrochemical cell for the indirect electrolytic destruction of organic wastes. (Adapted from Ibanez, 1997, and Ibanez, 2006).

introduction into the Ca(OH)$_2$ solution, a piece of Cu wire may be inserted in the pipet stem.

Heat a water bath in a 50-mL beaker to 40–50°C and maintain the temperature in this range during the entire experiment. Place the bulb of the pipet in the water bath. Connect the wires with alligator clips to the positive (Pb wire) and negative (paper clip) ends of a DC power source (e.g., a 9 V battery, an AC/DC adapter or an adjustable DC power source). If a power source other than a battery is used, turn it on at this point (adjust the potential to at least 3V). Should the voltage fall considerably below this value (like in the case of a used battery), there would not be enough driving force to carry out the desired reaction and another battery would be required.

Once the power source is turned on, bubbles are observed at both electrodes. Water is being reduced to hydrogen at the cathode, whereas at the anode, water is being oxidized to oxygen gas and simultaneously Co^{2+} is being oxidized to Co^{3+}. Some of the organic compound is also being directly oxidized at this electrode. In turn, the Co^{3+} ions thus produced oxidize the organic compound in the solution to CO$_2$. When the CO$_2$ comes in contact with the freshly prepared Ca(OH)$_2$ solution, insoluble CaCO$_3$ is formed and the clear Ca(OH)$_2$ solution becomes milky. When all the organic compound has been destroyed (i.e. after some 10–30 min., depending on the amount of organic added, the number of carbons in it and the applied voltage), the solution turns gray-light purple since the Co^{3+} ions keep on being formed but are no longer used up to oxidize the waste. This is an indication of the end of the reaction. The Co^{3+} can also oxidize water, although this reaction is considerably slower. At this point, the liquid can be squeezed out of the pipet into a small beaker and some more surrogate organic waste can be added to it (while still hot) to be oxidized. An immediate change in color (to the initial pink coloration) is then observed. (Add for example one drop of glycerin or ethanol. A few drops of acetic acid can also be added, although the color change is not so dramatic). To test for the presence of CaCO$_3$ in the collection tube, add to it a few drops of dil. H$_2$SO$_4$ or dil. CH$_3$COOH that will dissolve the precipitate and produce CO$_2$ again.

At the end of the experiment, return all portions—both used and unused—of the Co(II) solution to the instructor since this can be later reused (previous oxidation of any residual organic content by the above procedure). In addition, a small amount of Pb may be present in this electrolyte as a consequence of the anodic process and thus a reuse of this solution will be a more environmentally sound procedure. Likewise, return the Pb electrodes for reuse.

Name_____Section_____Date_____

Instructor_____Partner_____

PRELABORATORY REPORT SHEET—EXPERIMENT 13

Experiment Title _____

Objectives

Flow sheet of procedure

Waste containment and recycling procedure

PRELABORATORY QUESTIONS AND PROBLEMS

1. Compare the following redox mediators: Ag(II/I), Co(III/II) and Fe(III/II) in terms of a) their oxidizing power (i.e., compare their standard potentials), b) their cost *per mole of electrons* removed (assume that the most common nitrate salts are used for the initial, low oxidation state of each redox couple), and c) the possibility of using them with chlorinated organic wastes.
(*Hint:* costs can be looked up in any commercial catalog of chemicals or in the Chemical Marketing Reporter).

2. When should an indirect method be considered for the electrochemical oxidation of a waste?
3. Write down the (balanced) equation that describes the reaction between calcium hydroxide and carbon dioxide.

Additional Related Projects

- Use the Ag(II/I) or Fe(III/II) couples instead of the Co(III/II) to oxidize a surrogate organic waste. Exercise caution in the handling of Ag(I) solutions, because dark spots are easily produced on skin, tables, books, etc. In addition, a very acidic medium is required in the Ag process and should be handled with great care.

- Make this Co(III/II) experiment quantitative by performing it at the macroscale level and by carefully measuring the weight or volume of the nonvolatile organic compound added to the mediator solution. Filter, dry, and weigh the $CaCO_3$ precipitate obtained after the run, calculate the amount of CO_2 produced, and estimate the percent of destruction obtained. When CO_2 is the final product, this last process is called *mineralization*.

Name_____Section_____Date_____

Instructor_____Partner_____

LABORATORY REPORT SHEET—EXPERIMENT 13

Observations

1. Color of the initial electrolytic medium   _____

2. Temperature of the solution (initial)   _____°C

3. Number of *drops* of surrogate organic pollutant added   _____drops

4. Weight of $Ca(OH)_2$ used for the preparation of
   its saturated solution   _____g

5. Color of the final electrolytic medium   _____

7. Temperature of the solution (final)   _____°C

8. Time elapsed for complete reaction   _____min

POSTLABORATORY PROBLEMS AND QUESTIONS

1. Calculate the number of moles of surrogate organic waste added to your electrolytic medium.

2. Calculate the number of moles of CO_2 theoretically produced (assuming 100% yield).

*3. The oxidation of isopropanol with Co^{3+} in aqueous solution yields CO_2 as follows:

$$C_3H_7OH + H_2O_{(1)} + Co^{3+}$$
$$\rightarrow CO_{2(g)} + H^+ + Co^{2+}$$

Balance the equation. (*Hint:* do this procedure based on the changes in oxidation numbers).

4. Chlorinated organic compounds have varying degrees of toxicity. Some of them can be converted into benign inorganic species as follows:

$$C_3H_6(OH)Cl_{(1)} + H_2O_{(1)} + Co^{3+}$$
$$\rightarrow CO_{2(g)} + H^+ + Co^{2+} + Cl^-$$

Balance the equation (see the hint in the previous problem).

Student Comments and Suggestions

Literature References

Bersier, P. M.; Carlsson, L.; Bersier, J., "Electrochemistry for a Better Environment," in: *Topics in Current*

_____
* Answer in this book's webpage at www.springer.com

Chemistry, Vol. 170; Springer-Verlag: Berlin-Heidelberg, 1994.

Farmer, J. C.; Wang, F. T.; Lewis, P. R.; Summers, L. J. "Destruction of Chlorinated Organics by Co(III)-Mediated Electrochemical Oxidation," *J. Electrochem. Soc.* **1992**, *139*, 3025.

Farmer, J. C.; Wang, F. T.; Hawley-Fedder, R. A.; Lewis, P. R.; Summers, L. J.; Foiles, L. "Electrochemical Treatment of Mixed and Hazardous Wastes: Oxidation of Ethylene Glycol and Benzene by Silver," *J. Electrochem. Soc.* **1992**, *139*, 654.

Genders, J. D.; Weinberg, N., Eds. *Electrochemistry for a Cleaner Environment*; The Electrosynthesis Co.: E. Amherst, NJ, 1992.

Ibanez, J. G.; Singh, M. M.; Szafran, Z.; Pike, R. M. "Laboratory Experiments on Electrochemical Remediation of the Environment. Part 2: Microscale Indirect Electrolytic Destruction of Organic Wastes," *J. Chem. Educ.* **1997**, *74*, 1449–1450.

Ibanez, J. G. "Electrochemistry for Environmental Remediation: Laboratory Experiments," *Educ. Quim.* (Mexico, in English), **2006**, *17*, 274–278.

Kalvoda, R.; Parsons, R., Eds. *Electrochemistry in Research and Development*; Plenum Press: NY, 1985.

Pletcher, D.; Walsh, F., *Industrial Electrochemistry*, 2nd. ed.; Chapman and Hall: London, 1990.

Pletcher, D.; Weinberg, N. L. "The Green Potential of Electrochemistry. Part 1: The Fundamentals," *Chem. Eng.* 98 (Aug. **1992**).

Pletcher, D.; Weinberg, N. L. "The Green Potential of Electrochemistry. Part 2: The Applications," *Chem. Eng.* 132 (Nov. **1992**).

Rajeshwar, K.; Ibanez, J. G.; Swain, G. "Electrochemistry and the Environment," *J. Appl. Electrochem.* **1994**, *24*, 1077–1091.

Rajeshwar, K.; Ibanez, J.G. *Environmental Electrochemistry*; Academic Press: San Diego, CA (1997).

Sequeira, C. A. C., Ed. *Environmentally Oriented Electrochemistry*; Studies in Environmental Science #59, Elsevier: Amsterdam, 1994.

Steele, D. F. "Electrochemistry and Waste Disposal," *Chem. Brit.* **1991**, *27*, 915–918.

Tatapudi, P.; Fenton, J. R., "Electrolytic Processes for Pollution Treatment and Pollution Prevention," in: Gerischer, H.; Tobias, C. W., Eds. *Advances in Electrochemical Science and Engineering*, Vol. 4; VCH: Weinheim, 1995.

Walsh, F.; Mills, G. "Electrochemical Methods for Pollution Control," *Chem. Technol. Europe*, 13 (April–May **1994**).

Walsh, F.; Mills, G. "Electrochemical Techniques for a Cleaner Environment," *Chem. and Ind.* **1993**, *15*, 576–580.

Experiment 14
Electrochemical Treatment of Polluted Soils

Reference Chapters: 5, 10

Objectives

After performing this experiment, the student shall be able to:

- Understand and apply an electrochemical method for the removal of pollutants from a soil matrix.
- Interpret the pH gradients observed during electrokinetic phenomena.
- Predict the direction of the electroosmotic flow associated with electrokinetic phenomena.
- Understand the role of the anode and the cathode of an electrochemical cell.

Introduction

As discussed in Chapter 10, for the remediation of a piece of land suitable anodes and cathodes can be strategically placed in the (wetted) ground and an electric field from a DC source applied. As presented in Chapter 6, soil particles in contact with aqueous media are frequently negatively charged due to adsorption phenomena and lattice imperfections. Then, a double layer naturally forms inside a charged soil pore when cations from the liquid tend to neutralize this charge, and the outer layer of the liquid typically becomes positively charged as a result. The applied electric field produces a movement of this outer layer, and a drag interaction between this layer and its bulk inside the soil pore results. The liquid then moves along the potential field to wells where it can be collected and removed. This phenomenon is called *electroosmotic transport*.

In addition, electrochemical reactions at the electrodes produce $H_{2(g)}$ and $OH^-_{(aq)}$ (cathode) and $O_{2(g)}$ and $H^+_{(aq)}$ (anode), due to the electrolysis of water. These charged species (H^+ and OH^-), along with other ions encountered in the medium, are attracted to the oppositely charged electrodes and migrate, creating an acidic and a basic front, respectively. The movement of these fronts is aided by concentration gradients that promote diffusion.

The combined effects of all the phenomena involved (i.e., electric, chemical, and hydraulic potentials) is known as *electrokinetic remediation* or *electrokinetic processing* of soils. The importance and variety of electrochemical applications in the pollutant treatment and remediation areas is discussed more in depth in Section 10.1.

Three phenomena occurring during the electrokinetic processing of soils will be analyzed in this experiment by treating simulated contaminated soil (e.g., a clay or silica) with an electric field: a) the migration of ionic pollutants, b) the production and movement of acidic and basic fronts, and c) the movement of water under flow-direction control.

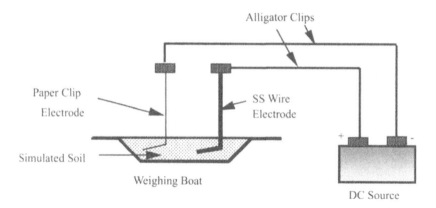

FIGURE 1. Microscale electrokinetic processing of soils. (Adapted from Ibanez, 1998).

Experimental Procedure

Estimated time to complete the experiment: 1.5 h

Materials	Reagents
3 weighing dishes	silica gel
1 2-mL graduated pipet	0.1 M Na_2SO_4
1 glass rod	isopropyl alcohol
2 stainless steel wires	thymol blue indicator
2 paper clips	phenolphthalein indicator
1 9-V battery, or an AC/DC	0.1 M $CuSO_4$
converter	0.01 M $K_2Cr_2O_7$
3 Beral pipets	D.I. water
pH indicator	

a) Demonstration of the electroosmotic effect

Prepare a simulated soil sample by weighing approximately 1.5 g of fine silica gel (e.g., of the type used for preparative layer chromatography) in a small plastic weighing dish or a similar container. Similar dishes will be used as soil containers throughout this experiment. Add drop wise (and evenly) 2–3 mL of 0.1 M Na_2SO_4 throughout the entire surface of the simulated soil as a supporting electrolyte to increase electrical conductivity. For more dramatic effects, kaolin (hydrated aluminum silicate) may be used; in this case, add 1 mL of the 0.1 M Na_2SO_4 solution. Stir the resulting mixture with a glass rod or a paper clip to form a homogeneous paste. Then, insert a piece of stainless steel (SS) wire as the anode in the simulated soil on one side of the dish (Pt or Pb also work well; a common paper clip may be used if SS, Pt, or Pb were not available). If Pb were used, see the precautions in

Experiment 13. Insert a common paper clip as the cathode on the opposite side, as shown in Figure 1. Cathodes made of iron, stainless steel and graphite also work well.

Connect the SS electrode to the **positive** terminal of a DC power source (e.g., a 9 V battery) and the paper clip to the **negative** terminal with alligator clips. Within a few minutes, water accumulates at the surface around the cathode due to the electroosmotic movement of water. To make the effect more dramatic, some drops of pH indicator can be added at this point. Because the water coming out of the cathode is basic due to the production of OH^- during water electrolysis, an indicator such as phenolphthalein yields a strongly red colored solution here (other suitable indicators may be used as well). At the same time, the area surrounding the anode will appear dryer.

b) Demonstration of the production and movement of acidic and basic fronts

Weigh approximately 1.5 g of fine silica gel in a small weighing dish. Insert the same electrodes as in the previous step in the simulated soil, as shown in Figure 1. Add drop wise (and evenly) 2–3 mL of 0.1 M Na_2SO_4 throughout the entire simulated soil. In the same manner, add enough drops of a thymol blue indicator solution (1 mg of indicator dissolved in 1 mL isopropyl alcohol, then diluted to 10 mL with water) so the entire surface is slightly yellow. Connect the electrodes as described above. Within a few minutes, a red spot should appear near the anode and a blue color around the cathode. [It may be necessary at this point to add more indicator to improve the visibility of the colors. The color changes (end points) of thymol blue are pH 1.2/red; 2.8/yellow; 9.2/ blue]. If time permits, wait a few more minutes and observe the displacement of the colored spots, which shows the movement of the acid and basic fronts.

c) Demonstration of metal ion migration

The components of a mixture of positive Cu^{2+} and negative $Cr_2O_7^{2-}$ ions will be separated by migration under an electric field. **Caution: Cr(VI) compounds are highly toxic and can be carcinogenic. This part should be performed by the instructor as a demonstration.** Weigh approximately 1 g of fine silica gel in a small weighing dish. Add drop wise (and evenly) 2 mL of a green solution 0.1 M in $CuSO_4$ and 0.01 M in $K_2Cr_2O_7$ throughout the entire simulated soil. Place the electrodes and connect them to the DC source as mentioned above. Within a few minutes, some bubbling will be noted at both electrodes due to the evolution of H_2 and O_2 (if no activity is observed, add 1 mL of the sodium sulfate solution described above). In addition, a yellow-orange spot will be noted at the anode and a blue spot at the cathode, thus showing the electromigration of the anion and cation, respectively.

This demonstration can also be done using $KMnO_4$ in place of $K_2Cr_2O_7$ in order to lower the toxicity of the sample. However, the experiment will take more time and the results are not as spectacular. Collect the chromium-containing residues in an appropriate container and follow local regulations for disposal.

Name _____ Section _____ Date _____

Instructor _____ Partner _____

PRELABORATORY REPORT SHEET—EXPERIMENT 14

Objectives

Flow sheet of procedure

Waste containment and recycling procedure

PRELABORATORY QUESTIONS AND PROBLEMS

1. Find three examples of substances that have been treated in real situations by electrokinetic remediation.

2. Biological methods are recommended as the first choice in many soil remediation scenarios due in part to their low cost. Name at least two situations where, in general, electrochemical methods are recommended instead.

3. The production of a basic front that originates at the cathode in a soil electrokinetic remediation site produces, as a side effect, the precipitation of a metal ion front, M^{n+} that is attracted and travels toward the cathode. In other words, the OH^- and M^{n+} ions meet at a certain point in the field and an insoluble metal hydroxide, $M(OH)_n$, precipitates as a result. (This phenomenon is similar to the analytical technique called *isoelectric focusing*). The end result is that the metal becomes concentrated ("focused") in a narrow band, which facilitates its removal. Assuming that the OH^- ions travel twice as fast as the monovalent metal ion, M^+ draw a one-dimensional sketch simulating the remediation site (the two extremes are the two electrodes placed for remediation), and mark the point where the *focusing* should result.

Additional Related Projects

- Try insoluble household powders instead of kaolin or silica gel and repeat this experiment (e.g., try talc, lime, powdered chalk, flour). Note your observations and propose an explanation for each result observed. (Hint: are these powders reacting with the fronts produced by the applied electric field, or are they inert towards them?)

- During the experiment on the movement of acidic and basic fronts, the rate of displacement of the red and blue spots can be measured (in cm/s) and plotted. Because protons have a higher transport number, one would expect their rate to be higher (after an initial induction period). (See Ibanez-Velasco, 2005).

Name _____ Section _____ Date _____

Instructor _____ Partner _____

LABORATORY REPORT SHEET—EXPERIMENT 14

a) Demonstration of the electroosmotic effect

Cathode material   _____

Anode material   _____

Power source   _____

Applied voltage   _____V

Observations:

Effect around the anode   _____

Effect around the cathode   _____

Indicator used   _____

b) Demonstration of the production and movement of acidic and basic fronts

Cathode material   _____

Anode material   _____

Power source   _____

Applied voltage   _____V

Observations:

Effect around the anode   _____

Effect around the cathode   _____

Indicator used   _____

c) Demonstration of metal ion migration

Cathode material   _____

Anode material   _____

Power source   _____

Applied voltage   _____V

Colored anion used   _____

Colored cation used   _____

Observations:

Effect around the anode   _____

Effect around the cathode   _____

POSTLABORATORY PROBLEMS AND QUESTIONS

1. Write the two chemical equations responsible for the observed production of an acidic and a basic front at the anode and cathode, respectively.

2. Solution movement inside a pore composed of charged particles can be described in the x direction by a total material influx (q_{tc}) equation:

$$q_{tc} = [-D_j(\partial C_j/\partial x) \\ + (nF/RT)D_j C_j (\partial E/\partial x) + V_x C_j]$$

where C_j = concentration of solute, D_j = diffusion coefficient of the charged particle, n = charge on the ion, F = Faraday's constant, R = universal gas constant, E = electrical potential, V_x = average seepage velocity. The three terms in this equation represent the following components of the flow: convection in a soil pore (also called advection), diffusion, and migration. Assign each term to the flow component that it represents.

3. Why is it possible to remove non-polar components from a polluted soil by electrokinetic remediation?

Student Comments and Suggestions

Literature References

Acar, Y. B.; Li, H.; Gale, R. J. "Phenol Removal from Kaolinite by Electrokinetics," *J. Geotech. Eng.* **1992**, *118*, 1837–1851.

Acar, Y. B.; Alshawabkeh, A. N. "Principles of Electrokinetic Remediation", *Environ. Sci. Tech.* **1993**, *27*, 2638–2647.

Cabrera-Guzman, D.; Swartzbaugh, J. T.; Weisman, A. W. "The Use of Electrokinetics for Hazardous Waste Site Remediation," *J. Air Waste Manag. Assoc.* **1990**, *40*, 1670–1676.

Hamed, J.; Acar, Y. B.; Gale, R. J. "Pb(II) Removal from Kaolinite by Electrokinetics," *J. Geotech. Eng.* **1991**, *117*, 241–271.

Ibanez, J. G.; Singh, M. M.; Szafran, Z.; Pike, R. M. "Laboratory Experiments on Electrochemical Remediation of the Environment. Part 3. Microscale Electrokinetic Processing of Soils," *J. Chem. Educ.* **1998**, *75*, 634–635.

Ibanez, J. G.; Takimoto, M. M.; Vasquez, R. C.; Basak, S.; Myung, N.; Rajeshwar, K. "Laboratory Experiments on Electrochemical Remediation of the Environment: Electrocoagulation of Oily Wastewater," *J. Chem. Educ.* **1995**, *72*, 1050–1052.

Ibanez-Velasco, G.; Velasco-Herrejon, P. "Electrolysis Applied To Environmental Clean-Up," *Prax. Naturwiss. Chem. Sch.* (Germany, in English), **2005**, *54 (4)* 34–36.

Krishnan, R.; Parker, H. W.; Tock, R. W. "Electrode Assisted Soil Washing," *J. Haz. Mat.* **1996**, *48*, 111–119.

Lageman, R. "Electroreclamation," *Environ. Sci. Tech.* **1993**, *27*, 2648–2650.

Little, J. G. "A Simple Demonstration of Ion Migration," *J. Chem. Educ.* **1990**, *67*, 1063–1064.

Mattson, E. D.; Lindgren, E. R.; "Electrokinetic Extraction of Chromate from Unsaturated Soils," Chapter 2 in: Tedder, D. W.; Pohland, F. G., Eds. *Emerging Technologies in Hazardous Waste Management-V*; American Chemical Society Symposium Series #607: Washington, 1995.

Probstein, R. F.; Hicks, R. E. "Removal of Contaminants from Soils by Electric Fields," *Science* **1993**, *260*, 468–503.

Rajeshwar, K.; Ibanez, J. G., *Environmental Electrochemistry*; Academic Press: San Diego, 1997.

Rajeshwar, K.; Ibanez, J. G.; Swain, G. "Electrochemistry and the Environment," *J. Appl. Electrochem.* **1994**, *24*, 1077–1091.

Segall, B. A.; Bruell, C. J. "Electroosmotic Contaminant-Removal Processes," *J. Environ. Eng.* **1992**, *118 (1)*, 84–100.

Shakhashiri, B. Z. *Chemical Demonstrations: A Handbook for Teachers of Chemistry*, Vol. 4; The University of Wisconsin Press: Madison, WI. 1992. p. 150.

Shapiro, A. P.; Probstein, R. F. "Removal of Contaminants from Saturated Clay by Electroosmosis," *Environ. Sci. Tech.* **1993**, *27*, 283–291.

Smollen, M.; Kafaar, A. "Charged Up Solids Aid Dewatering in Filter Belt Press," *Wat. Environ. Technol.* **1995**, *7 (11)*, 13–14.

Thornton, R. F.; Shaphiro, A. P. "Modeling and Economic Analysis of In Situ Remediation of Cr(VI)-Contaminated Soil by Electromigration," Chapter 4 in: Tedder, D. W.; Pohland, F. G., Eds. *Emerging Technologies in Hazardous Waste Management-V*; American Chemical Society Symposium Series #607: Washington, 1995.

Trombly, J. "Electrochemical Remediation Takes to the Field," *Environ. Sci. Tech.* **1994**, *28*, 289A–291A.

Experiment 15
Removal of Nitric Oxide by Complex Formation

Reference Chapters: 8, 10

Objectives

After performing this experiment, the student shall be able to:

- Prepare the gaseous pollutant NO.
- Understand the acid–base behavior of NO and its possible environmental impact.
- Produce a metal chelate.
- Understand and apply a complexation method for the removal of an insoluble gas.
- Regenerate the chelate for further use.

Introduction

As discussed in Chapter 8 of the companion book, the emission of nitric oxide (NO) is problematic due to

- its contribution to ozone depletion in the ionosphere (i.e., $NO + O_3 \rightarrow NO_2 + O_2$)
- the production of smog in the troposphere
- its proclivity to attach to hemoglobin (preventing dioxygen transport)
- its role as a precursor of acid rain
- its high reactivity (it is a radical)

NO is thermodynamically unstable at room temperature and pressure. As soon as it comes in contact with dioxygen, it is oxidized to nitrogen dioxide (a toxic, brown acidic gas).

Many processes have been proposed and used for NO removal from combustion gases. Dry processes include its reduction with NH_3 and N_2H_4. Wet processes are often aimed at the simultaneous removal of SO_2 and NO. However, owing to the low solubility of NO in water (1.25×10^{-3} M atm^{-1} at 50°C), chemical reactions in the gas phase or in aqueous media are needed to solubilize it. Practically all of the NO wet scrubbing methods are based on its oxidation (e.g., with H_2O_2), reduction (e.g., with SO_3^{2-}), or complexation (e.g., with aminopolycarboxylates). In the present experiment we will focus on this last approach.

Complex formation removes gaseous pollutants when these can be used as ligands for selected metal ions, typically in aqueous solution. The complexes thus formed can be chemically or electrochemically oxidized or reduced, whereby the pollutant (acting as a ligand) is destroyed to produce less harmful species, and the metal ion becomes ready to reinitiate the cycle.

Removal of NO by complexation can be as high as 70–80%, with typical enhancement factors of 100–500 as compared to its absorption in pure water. Several iron thiochelates (e.g., with 2,3-dimercapto-1-propanesulfonate) are particularly effective for its removal. Iron aminopolycarboxylate chelates (e.g., Fe–EDTA) are also effective to a satisfactory degree.

An Fe–EDTA chelate will be used here due to the wider availability and lower cost of aminopolycarboxylates as compared to thiochelates. In addition, EDTA is a strong complexing agent for many metal ions, although unfortunately this same property makes it a persistent pollutant.

The complexing ability of metal chelates can be selective for certain oxidation states. For example, [Fe(II)EDTA] forms with NO the corresponding (mono)nitrosyl complex [Fe(II)NO(EDTA)], whereas the Fe(III) chelate cannot complex it. For this reason, a practical NO scrubbing system must

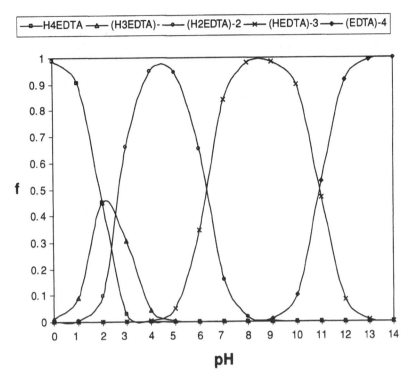

FIGURE 1. EDTA speciation with pH.

include the Fe(III) to Fe(II) reduction step. Reducing agents for this step include SO_3^{2-} and HSO_3^- ions formed by the dissolution of SO_2 (which frequently accompanies NO as an emission gas). Other reducing agents suitable for this purpose are $S_2O_4^{2-}$, S^{2-}, ascorbic acid, iron metal, glyoxal, and electrochemical reduction. The NO-reduction products depend on the reduction method used and include NH_4^+, NH_3, N_2, N_2O, $NH_2(SO_3H)$, and $HON(SO_3)_2^{2-}$.

Solution pH can be of paramount importance to have an adequate ligand speciation (see Section 2.1). For example, EDTA can typically be present as five different chemical species depending on the pH of the medium: H_4EDTA, H_3EDTA^-, H_2EDTA^{2-}, $HEDTA^{3-}$, and $EDTA^{4-}$ (see Figure 1). Its complexes can then be anionic, cationic, or neutral, and this can have dramatic effects on their properties and behavior. A case in point is the EDTA complexation of Fe^{3+}, in which the anionic complex with $EDTA^{4-}$, $[FeEDTA]^-$, is some 20 orders of magnitude more stable than the corresponding neutral complex with $HEDTA^{3-}$, $[FeHEDTA]$. (For ligand speciation calculations one can use the chemical-species distribution algorithm described in Section 2.1 and developed by Rojas, 1995).

In the experiment presented below, the [Fe(II)EDTA] complex will be prepared and used to demonstrate the reactive dissolution of an insoluble polluting gas (NO) by complex formation. The NO in the resulting complex is then reduced and the [Fe(II)EDTA] complex is regenerated.

Experimental Procedure

Estimated time to complete the experiment: 2 h

Materials	Reagents
2 5-mL vials	Fe(II)EDTA
2 20-mL filter flasks	KNO_2
2 10-mL syringes	1 M H_2SO_4
1 1-mL graduated pipet	1 M KOH
1 2-mL graduated pipet	$FeSO_4 \cdot 7H_2O$
2 10-mL beakers	Na_2H_2EDTA
3 25-mL Erlenmeyer flasks	$Na_2S_2O_4$
2 plastic septa	nitrogen gas
1 well plate	D. I. water
1 plastic cap	
1 spectrophotometer cell	
1 spectrophotometer	
20 cm of rubber tubing	
1 magnetic stirrer	

This experiment consists of five steps: a) Preparation of NO, b) preparation of a basic trap, c) preparation of [Fe(II)EDTA], d) removal of NO by [Fe(II)EDTA], and e) reduction of the corresponding iron nitrosyl complex to regenerate the chelate.

This sequence is conveniently performed, for example, in two 5-mL vials and a 20-mL filter flask (see Ibanez, 2000). The vials need to be equipped with septa and connected to the filter flask by means of short pieces of plastic or rubber tubing (e.g., 3–5 mm in diameter). A well plate (with a well capacity of at least 3 mL) equipped with plastic caps and tubes for every well to convey the produced gases from one well to the next, can also be used (at least four wells are required).

a) Preparation of NO

NO can be conveniently prepared by the reaction represented in equation 1:

$$6NaNO_{2(s)} + 3H_2SO_4 \rightarrow 4NO_{(g)} + 2H_2O_{(1)}$$
$$+ 3Na_2SO_4 + 2HNO_3$$
$$(1)$$

Place 1 mL of deionized or distilled water and about 20 mg of KNO_2 or $NaNO_2$ in one of the 5-mL vials (also called the *preparation* vial). Cap it with a two-hole rubber septum equipped with two pieces of tubing inserted through it, one of which is connected to a nitrogen gas tank, and the other to the filter flask, as shown in Figure 2. Be ready to add approximately 0.2 mL of 1 M H_2SO_4 with a syringe through the rubber septum (secure a tight fit to prevent the escape of gas), but do not add the acid until the entire set-up is ready (i.e., after step c). **Caution: NO gas may be irritating to the eyes, throat, and nose, and it is known to have other biological effects including increased vasodilatation. NO_2 is toxic.**

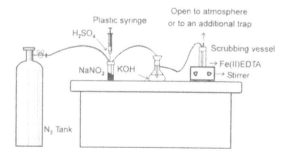

For these reasons, it is strongly recommended that the experiment be performed under a fume hood in a well-ventilated area.

b) Preparation of a basic trap

A KOH trap is prepared by placing 10 mL of 1 M KOH in the filter flask (also called the *basic trap*). Connect the exit tube from the first vial to this flask and dip the tube in the basic solution. Connect the exit tube from this flask to the second 5-mL vial in the same manner so as to transport the NO into the chelate solution that will be prepared in this vial, see Figure 2.

c) Preparation of the [Fe(II)EDTA] chelate

Place 5 mL of distilled or deionized water in a 10-mL beaker. While stirring, add the necessary amount of an Fe(II) salt (for example, $FeSO_4 \cdot 7H_2O$) to form a 0.3 M Fe^{2+} solution. Transfer enough of this solution to a spectrophotometer cell and take its absorption spectrum. Return the solution into the beaker.

Then, add enough EDTA (for example, its disodium salt) to form the 1:1 iron chelate and stir. Excess EDTA can be used to ensure complete chelate formation. Avoid low pH values because the EDTA may form its tetraprotonated, insoluble form or else the positively charged complex, $[Fe(H_3EDTA)]^+$, as deduced from Figure 1. Both forms are unsuitable for the present experiment.

Transfer 1–2 mL of this freshly formed chelate solution into a spectrophotometric cell and obtain a scan in the visible region with a UV-VIS spectrophotometer. Transfer another 2 mL of this chelate solution to the second 5-mL vial (also called the *scrubbing vessel*). Make sure that all the tubes and the syringe are now connected. Nitrogen gas can be used as the carrier gas. Initiate gentle nitrogen bubbling so as to carry the NO from the preparation well into the basic trap and into the scrubbing vessel.

[A simpler method consists of using a large syringe to inject air continuously to the preparation vial for this same purpose; a certain disadvantage in using this alternative approach is that some NO will be oxidized to NO_2, and some Fe(II) will be oxidized to Fe(III), but this may be offset by its simplicity because the use of a compressed gas is avoided].

Caution: Do not start producing NO gas until the entire set-up is ready, the iron chelate has been produced (see below), and the carrier gas is either flowing or ready to be pumped into the system.

Now, inject the acid into the preparation vial to initiate NO formation as described in a).

d) Removal of NO by [Fe(II)EDTA], producing [Fe(II)NO(EDTA)]

After a few minutes of bubbling the carrier gas containing NO through the scrubbing solution, a color change will be observed (from pale yellow to olive green) due to the formation of [Fe(II)NO(EDTA)]. At this point, stop the production of NO by opening the preparation well and flushing it with running water. The small amount of residual nitrite can be flushed down the drain. Transfer 1–2 mL of the green solution from the scrubbing well (containing the nitrosyl complex) to a spectrophotometric cell, and take its visible absorption spectrum.

e) Reduction of the [Fe(II)NO(EDTA)] complex to regenerate the [Fe(II)EDTA] chelate

In order to reduce the NO in the nitrosyl complex and regenerate the iron chelate, return the solution from the spectrophotometer cuvette into the scrubbing vessel, and add some 20 mg of dithionite powder (sodium hydrosulfite, $Na_2S_2O_4$) to it. Stir. (Note: if not enough dithionite were added, the mixture becomes red with time!) After some time, a color change back to the original pale-yellow is observed, signaling the destruction of the NO complex and the regeneration of the [Fe(II)EDTA] chelate. It is a good practice to set aside a small amount of the untreated chelate for color comparison. (Sodium sulfite can also be used for this reduction step, although the color change will be somewhat less spectacular). Use some of this reduced solution to obtain its visible spectrum.

The three spectra (i.e., before chelation, after chelation, and after reduction) can be now compared. An intense absorption band is observed in the second spectrum, which is absent in the other two. This demonstrates the formation and destruction of the NO complex.

Name_____Section_____Date_____

Instructor_____Partner_____

PRELABORATORY REPORT SHEET—EXPERIMENT 15

Objectives

Flowsheet of the procedure

Waste containment and recycling procedure

PRELABORATORY QUESTIONS AND PROBLEMS

*1. Propose at least five oxidants for NO based on Mn, Cl, or Cr compounds.

*2. Explain the complex-formation ability of NO by drawing its Lewis structure.

*3. a) Write the electronic structure of NO. (You may assume that it is the same as that of O_2, with one electron removed). Hint: Recall that in order to get the electronic configuration of homonuclear diatomic molecules (X_2) of the first short period of the Periodic Table, one assumes that the inner molecular orbitals formed by the $1s$ orbitals from both atoms are filled (this takes four electrons). The next higher energy orbitals are formed by the $2s$ and $2p$ orbitals. In dioxygen, the resulting orbital arrangement is: $(\sigma_1)(\sigma_2^*)(\sigma_3)(\pi_1)(\pi_1)(\pi_2^*)(\pi_2^*)(\sigma_4^*)$

b) Based on your response to a), is NO diamagnetic or paramagnetic?

c) Would you predict the nitrogen-oxygen bond in the NO^+ ion to be stronger or weaker than that in NO?

Additional Related Projects

- Produce NO and capture it by reactive absorption involving its oxidation (e.g., with H_2O_2) or its reduction (e.g., with SO_3^{2-}). Test for the absence of NO in the gas exiting the selected scrubbing solution by showing the absence of the green color in a [Fe(II)EDTA] solution prepared as described above (step c).
- Produce NO and capture it by absorption in an aqueous iron (II) sulfate solution. This forms a distinctively brownish nitroso iron (II) sulfate.
- Roussin's black salt results from the action of NO upon an iron (II) sulfate/potassium thiosulfate solution. To prepare this solution, dissolve 5 g of $K_2S_2O_3$ in 5 mL of distilled water. Then, add 77 mg of $FeSO_4 \cdot 7H_2O$ and bubble NO (prepared as described above) through it. After drying the precipitate that forms, blackish-brown crystals should be observable under a microscope. This is a tetrahedral complex. (Generally speaking, tetrahedral complexes tend to have more intense absorption bands than the corresponding octahedral ones).

_____
*Answer in this book's webpage at www.springer.com

Name_____Section_____Date_____

Instructor_____Partner_____

LABORATORY REPORT SHEET—EXPERIMENT 15

a) Preparation of NO

Most likely, you prepared NO with A, B and C. Fill-in the blanks:

Formula of A   _____

Formula of B   _____

Formula of C   _____

Amount of A used   _____ (mg or mL)

Amount of B used   _____ (mg or mL)

Amount of C used   _____ (mg or mL)

b) Purification of NO

Was NO purified by passing the gases produced and the carrier gas through a trap previous to the scrubbing vessel? If affirmative, what was the composition of the trap?

c) Scrubbing the NO

Amount of Fe(II) added   _____ mg

Amount of EDTA added   _____ mg

Condensed formula of the EDTA used   _____

Color of the [Fe(II)EDTA] complex   _____

Color of the [Fe(II)(NO)(EDTA)] complex   _____

d) Regeneration of the [Fe(II)EDTA]

Reducing agent used (formula)   _____

Amount of reducing agent used   _____ mg

e) Analysis

If the VIS absorption spectra of the solutions of Fe(II), [Fe(II)EDTA], and [Fe(II)(NO)(EDTA)] were taken, insert here a copy and interpret qualitatively the differences observed. If the spectrum after reduction was taken, also insert it here and compare with the others.

f) Additional Tests

Was any additional test performed for the presence of NO? If affirmative, describe your results.

POSTLABORATORY PROBLEMS AND QUESTIONS

*1. Removal of insoluble NO from a flue-gas stream can be accomplished by scrubbing with a solution containing Fe(II)EDTA (as in the present experiment). The unavoidable presence of O_2 produces the oxidation of Fe(II)EDTA to Fe(III)EDTA, which is unreactive toward NO and thus renders the process useless. Consequently, the reduction of Fe(III)EDTA is a key step toward the success of the global process. To this end, bisulfite can be used as a reducing agent as follows (L symbolizes the EDTA ligand):

$$2Fe(III)L + 2HSO_3^- \rightarrow 2Fe(II)L + S_2O_6^{2-} + 2H^+ \tag{1}$$

This reaction is first order in [Fe(III)L] and order -1 (i.e., inverse order) in [Fe(II)L], as shown by the rate equation:

$$-\frac{d[Fe(III)L]}{dt} = \frac{k[Fe(III)L]}{[Fe(II)L]} \tag{2}$$

Reaction 1 may be assumed to be composed of the following **elementary** steps, each one characterized by its corresponding physicochemical parameters: a) K (equilibrium constant), and b) k (rate constant) as follows:

$$Fe(III)L + HSO_3^- = FeSO_3^+ + L + H^+ \quad K_3, k_3 \tag{3}$$

$$FeSO_3^+ = Fe(II) + {}^\bullet SO_3^- \quad\quad\quad K_4, k_4 \tag{4}$$

$$Fe(II) + L = Fe(II)L \quad\quad\quad\quad K_5, k_5 \tag{5}$$

$${}^\bullet SO_3^- + H^+ = HSO_3 \quad\quad\quad\quad K_6, k_6 \tag{6}$$

$$\text{(rate determining step)} \tag{6}$$

$$HSO_3 + HSO_3 = S_2O_6^{2-} + 2H^+ \quad K_7, k_7 \tag{7}$$

[Note that the HSO_3^- in equation 3 is *not* the same as the (neutral) HSO_3 in eqs. 6 and 7]. See Suchecki, 2005.

With this information:

a) Combine equations 3 through 7 to obtain equation 1 (i.e., use Hess' Law).
b) Write the mathematical expression for each equilibrium constant corresponding to reactions 3 to 7 (i.e., K_3, K_4, K_5, K_6, and K_7).
c) Derive the rate equation (equation 2) by assuming that the initial concentration of HSO_3^- is so large

that it essentially remains constant during the reaction, and knowing that the rate-determining step is reaction 6. Note that this **elementary** reaction obeys the rate equation:

$$-\frac{d[Fe(III)L]}{dt} = k_6[{}^\bullet SO_3^-][H^+]$$

*2. Some gases can be removed from waste gas streams by absorption-oxidation processes. For example, NO can be absorbed in a basic solution of a cobalt (III)–ethylenediamine complex, which acts as a catalyst for the oxidation of NO to NO_2, as shown in the following series of reactions (see Long, 2005):

$$Co(en)_3^{3+} + NO(g) + OH^- \rightarrow Co(en)_2(NO)OH^{2+} + en \tag{1}$$

$$2Co(en)_2(NO)OH^{2+} + O_{2(g)} \rightarrow 2Co(en)_2(NO_2)OH^{2+} \tag{2}$$

$$2Co(en)_2(NO_2)OH^{2+} + 4OH^- \rightarrow 2Co(en)_2(OH)_2^+ + NO_2^- + NO_3^- + H_2O_{(l)} \tag{3}$$

$$Co(en)_2(OH)_2^+ + en \rightarrow Co(en)_3^{3+} + 2OH^- \tag{4}$$

The global reaction is known to have the form:

$$2X + 2Y + Z \longrightarrow P + Q + R$$

Write this reaction using the true formulas of X, Y, Z, P, Q, and R.

Student Comments and Suggestions

Literature References

College, J. W.; Tseng, S. C.; MacKinney, D. "Process for Removing Sulfur Dioxide and Nitrogen Oxides from a Hot Gaseous Stream with Ferrous Chelate Regeneration," U.S. Patent 5,695,727 (Dec. 9, 1997).

Ealy, J. B.; Ealy, J. L. Jr. *Visualizing Chemistry. Investigations for Teachers*; American Chemical Society: Washington, DC, 1995. p 349.

___
* Answer in this book's website at www.springer.com

Gambardella, F.; Alberts, M. S.; Winkelman, J. G. M. "Experimental and Modeling Studies on the Absorption NO in Aqueous Ferrous EDTA Solutions," *Ind. Eng. Chem. Res.* **2005**, *44*, 4234–4242.

Ibanez, J. G.; Miranda-Treviño, J. C; Topete-Pastor, J; Garcia-Pintor, E. "Metal Complexes and the Environment: Microscale Experiments with Iron–EDTA Chelates," *Chem. Educ.* **2000**, *5*, 226–230.

Jones, K. *The Chemistry of Nitrogen*; Pergamon Press: Oxford, 1973. pp 323–335.

Juttner, K.; Kleifges, K.-H.; Juzeliunas, E. "Development of an Electrochemical Process for NO-Removal from Combustion Gases," in: Walton, C. W., Rudd, E. J., Eds. *Energy and Electrochemical Processing for a Cleaner Environment*; Electrochem. Soc. Proc.: New Jersey, 1998. Vol. 97-28, pp 439–450.

Juzeliunas, E.; Juttner, K. "Electrochemical Study of NO Conversion from Fe(II)-EDTA Complex on Pt Electrodes," *J. Electrochem. Soc.* **1998**, *145*, 53–58.

Kleifges, K.-H.; Kreysa, G.; Juttner, K. "An Indirect Electrochemical Process for the Removal of NO_x from Industrial Waste Gases," *J. Appl. Electrochem.* **1997**, *27*, 1012–1020.

Long, X.-L; Xiao, W.-D.; Yuan, W.-K. "Kinetics of Gas-Liquid Reaction between NO and $Co(en)_3^{3+}$," *Ind. Eng. Chem. Res.* **2005**, *44*, 4200–4205.

Rojas-Hernandez, A.; Ramirez, M. T.; Gonzalez, I.; Ibanez, J. G. "Predominance-Zone Diagrams in Solution Chemistry. Dismutation Processes in Two-Component Systems (M-L)," *J. Chem. Educ.* **1995**, *72*, 1099–1105.

Suchecki, T. T.; Mathews, B.; Kumazawa, Ho. "Kinetic Study of Ambient-Temperature Reduction of Fe(III)EDTA by $Na_2S_2O_4$," *Ind. Eng. Chem. Res.* **2005**, *44*, 4249–4253.

Tsai, S. S.; Bedell, S. A.; Kirby, L. H.; Zabcik, D. J. "Field Evaluation of Nitric Oxide Abatement with Ferrous Chelates," *Environ. Progr.* **1989**, *8*, 126–129.

Experiment 16
Photocatalytic Remediation of Pollutants

Reference Chapters: 8, 10

Objectives

After performing this experiment, the student shall be able to:

- Understand the concepts of band gap, electron and hole generation in a semiconductor.
- Observe the dual nature of the photocatalytic reactions.
- Photoreduce a metal ion in solution.
- Photooxidize a surrogate organic pollutant.
- Analyze the effect of the absence or presence of light, oxygen, and of a semiconductor on photocatalysis.

Introduction

As discussed in Chapters 3 and 10, light with energy higher than that of the band gap in a semiconductor (e.g., TiO_2) can excite an electron from its valence band to the conduction band. This process also creates a positive *hole* in the valence band. The electron and hole thus produced can reduce or oxidize different pollutants (e.g., organic compounds, inorganic species, metal ions, bacteria). A summary of the advantages, challenges, and proposed solutions in TiO_2 photocatalysis is given in Chapter 10. Even though a reduction and an oxidation always occur, only in a few cases are both the anodic and cathodic reactions advantageously utilized (see Colon, 2001 and Layman, 1995).

In the present experiment we describe a microscale experiment in which the *simultaneous* oxidation of an organic compound (ethanol or citric acid) and the reduction of a metal ion (Cu^{2+}) are photocatalytically performed in an aqueous slurry containing TiO_2 irradiated with UV light. The production of electrons (capable of reducing the metal ions) and of holes (capable of oxidizing the organic molecule) can be used for environmental clean-up.

Experimental Procedure

Estimated time to complete the experiment: a) Qualitative option: 1.5 h. b) Quantitative option: 2.5 h

Materials	Reagents
1 quartz UV- pencil lamp	0.04 M $CuSO_4$
1 100-mm long quartz tube	TiO_2 (preferably
2 rubber bands	Degussa P25)
1 100-mL beaker	D. I. water
1 1-mL graduated pipet	96% ethanol
1 2-mL graduated pipet	0.33 M citric acid
1 magnetic stirrer	$Ba(OH)_2$
2 Beral pipets	
1 stirring bar	
6 triangular bars	
6 one-hole rubber stoppers	
parafilm	
1 Pasteur pipet	
1 spectrophotometer	
1 spectrophotometer cell	
1 cardboard box to cover the UV source	
6 test tubes (5 cm long, 5 mm I.D.)	
1 black plastic bag	
1 glass capillary U-tube	

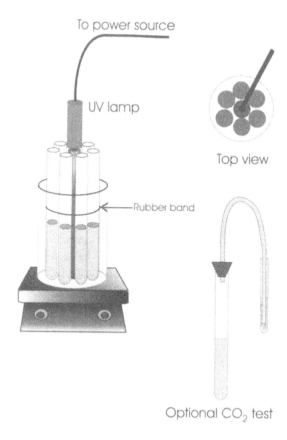

To power source

UV lamp

Top view

Rubber band

Optional CO$_2$ test

FIGURE 1. Experimental set-up. (Reproduced from the *Journal of Chemical Education* **2005**, *82*, 1549–1551, with permission).

Arrange a set-up consisting of a UV light source (e.g., a miniature UV-quartz pencil lamp, Spectroline model 11SC-1, Spectronics Corp.). If a miniature light source is used, it can be introduced into a quartz test tube (e.g., Ace Glass 13 mm OD, 100 mm long quartz tube without lip). Surround this tube with six quartz tubes containing different suspensions as described below, cover each one with parafilm, and secure the bundle with two rubber bands. Place the bundle inside a 100-mL beaker on top of a magnetic stirrer (see Figure 1). Add a small stirring bar to each of the six tubes (triangular bars of the type used with conical-bottom vials work well). Depending on your objectives, place the different solutions inside the test tubes according to the qualitative or quantitative procedures described below (or do both).

In the qualitative option, the test tubes will contain a Cu(II) solution, a TiO$_2$ suspension and some ethanol. One component will be missing from each test tube and the observed results interpreted accordingly. In the quantitative option, citric acid will be used instead of ethanol, and all the test tubes will have the same contents; irradiation time will be varied and the Cu(II) removed will be analyzed as a function of time.

Because the oxidation of the organic compound produces CO$_2$, you may desire to do a test for its production. To this end, instead of using parafilm, cap each quartz tube with a one-hole rubber stopper connected by a glass capillary U-tube to a small test tube (e.g., 5 cm long, 5 mm I.D.) half-filled with a saturated Ba(OH)$_2$ solution as a CO$_2$ trap (see Figure 1). Before initiating the UV illumination make sure that all six stirring bars are mixing well each solution. **Caution: UV light is dangerous. Avoid exposure of any part of the body (particularly the eyes), by placing a thick cardboard box or another suitable protective covering between the UV lamp and the experimenter. For added protection, cover the whole set-up with a black polyethylene bag. Copper (II) sulfate is a strong irritant to skin and mucous membranes. BaCO$_3$ and Ba(OH)$_2$ are poisonous. Overexposure to TiO$_2$ may cause slight lung fibrosis, and it is a potential occupational carcinogen.**

Because ambient dioxygen is a good electron trap, it should be removed from the suspensions since it competes as an electron acceptor with Cu(II). To this end, bubble nitrogen gas for a couple of minutes from a nitrogen tank into each test tube (except one, as described below) containing the prepared solutions and then seal each tube with parafilm.

1) *Qualitative procedure: Observation of the photocatalytic effect.*

With distilled or deionized water, prepare the following:

a) An aqueous slurry containing 100 mg of TiO$_2$ (e.g., Degussa P25) in 50 mL of water
b) A 0.04 M CuSO$_4$ solution
c) 96% ethanol

As instructed in Table 1, place the following components in the corresponding test tubes. TiO$_2$: add 2 mL of its suspension. Cu(II): add 2 mL of its solution. Ethanol: add 0.5 mL. Then, bubble nitrogen into the tubes, cover them with parafilm and proceed as instructed below.

TABLE 1. Components added to each test tube.

Tube #	TiO_2	Cu(II)	Ethanol	N_2	$h\nu$
1	–	✓	✓	✓	✓
2	✓	–	✓	✓	✓
3	✓	✓	–	✓	✓
4	✓	✓	✓	–	✓
5	✓	✓	✓	✓	–
6	✓	✓	✓	✓	✓

Note: To prevent light from reaching the tube #5, wrap it with Al foil, and subject it to the same conditions as the others.

Once all the components are in their respective test tubes, and these are all bundled and placed on site, cover the entire experimental set up with the box and bag described above and turn the UV light on. Then, wait approximately 10 min (this period of time works well for the UV lamp described above). After this time, turn it off, carefully observe the appearance of the resulting products in each test tube, write down these observations, and interpret them.

Interestingly enough, the reduction of Cu(II) to Cu(I) yields a reversible, insoluble purple complex with TiO_2 (see Foster, 1993). A very small amount of elemental copper is sometimes observed as a thin layer deposit on the walls of some quartz tubes (like a copper mirror) or even on the TiO_2 surface, which is further proof of the Cu(II) reduction. In addition, the production of a white precipitate in the small (optional) collection tubes confirms the production of CO_2 as a result of the photocatalytic oxidation of the organic molecule.

2) Quantitative procedure: Observation of Cu(II) removal as a function of time.

With distilled or deionized water, prepare the following:

a) An aqueous slurry containing 100 mg of TiO_2 (e.g., Degussa P25) in 50 mL of water
b) A 0.33 M citric acid solution
c) A 0.04 M $CuSO_4$ solution

Repeat the set-up preparation procedure exactly as described above, but this time add 2 mL of the TiO_2 suspension to *all six tubes*, as well as approximately 0.8 mL of the citric acid solution, and 2 mL of the $CuSO_4$ solution. Once all the components are in their respective test tubes, the nitrogen has been bubbled, and the tubes are covered with parafilm, bundled, and placed on site, cover the entire experimental set up with the box and bag described above. Turn the UV light on and allow irradiation

for approximately 10 min. Then turn it off and remove with a Pasteur pipet the contents of test tube 1. Cover the set-up again and turn the UV light on once more. For protection, move out of the irradiation area, quickly filter the contents of test tube 1 through a fine filter (for example, place the contents of the Pasteur pipet in a syringe equipped with a 0.45 μm syringe filter and push the contents out with the plunger), and analyze for the removal of Cu(II) with the aid of a colorimeter or spectrophotometer set at 810 nm. [You may want to see the absorption spectrum of Cu(II) in the experiment on *Metal Ion Recovery by Cementation* in this book; the Cu(II) as well as the Cu(II) + citric acid absorbance plots have their absorption maxima essentially at this same wavelength.] Note that a calibration plot is not needed since only the **normalized** absorbances (i.e., A_t/A_0) are needed so as to observe the removal of Cu(II). (Here, A_t is the absorbance at time t, and A_0 is the initial absorbance).

Repeat the same procedure with the remaining five test tubes, removing one at a time after every 10 min of irradiation. To show that light is required for this process to occur, a separate blank experiment may be done with a sample in a tube wrapped with Al foil, subject to the same conditions as the others, and then analyzed after a given period of time so as to compare it with an irradiated one.

A plot of the Cu(II) **normalized** absorbance decrease as a function of time will reveal the extent of removal. This is the result of a photocatalytic-adsorption combined effect, which can be tested separately if desired as described in the literature (see Bumpus, 1999).

Once the qualitative and/or the quantitative experiments are finished, students can bubble dioxygen from a tank into the purple suspension [containing $Cu(I)-TiO_2$] and observe within a few minutes the color change back to light-blue as a result of re-oxidation and dissolution of the $Cu^+_{(ads)}$ that reverts to $Cu^{2+}_{(aq)}$. Air can be used instead of dioxygen, but with less spectacular results. If desired, the photocatalyst may then be recovered by filtration with a 0.45 μm membrane filter. This photoredox cycling can be envisaged as an environmentally friendly metal removal/concentration cycle. This is why the filtration step described in the quantitative procedure must be done quickly so as to avoid metal re-dissolution, and this is an additional reason for removing air by bubbling nitrogen at the beginning of the experiment.

Name_____Section_____Date_____

Instructor_____Partner_____

PRELABORATORY REPORT SHEET—EXPERIMENT 16

Objectives

Flow sheet of procedure

Waste containment and recycling procedure

PRELABORATORY QUESTIONS AND PROBLEMS

*1. The possibility of using a semiconductor to cause a photocatalytic reaction is governed by the location of the valence and conduction bands (VB and CB, respectively) relative to the location of the desired reactions. For example, if one desires to split water into hydrogen plus oxygen, one needs to pay attention to the corresponding standard potential for the reduction of water into hydrogen (0.0 V vs the Standard Hydrogen Electrode, SHE) and that for the oxidation of water into oxygen (1.23 V). The electrons to be produced at the semiconductor must have a more negative standard potential than that required for the reduction of water, and the holes must be more positive so as to oxidize water. Assuming that the location of the bands in the selected semiconductor remains unchanged upon illumination, select the option that contains a combination of values suitable for water splitting. The locations of the VB and CB in a potential axis are given in volts, respectively.

a) (3.0, 0.2)

b) (1.0, 0.2)

c) (1.0, −0.8)

d) (2.0, −0.8)

e) (1.0, −2.0)

2. Is photocatalysis en **electrochemical** phenomenon? Explain.

*3. In a semiconductor-based **photosynthetic** system, light drives a non-spontaneous reaction. This is called a *thermodynamically up-hill* reaction. On the other hand, if the reaction is thermodynamically favored (i.e., *thermodynamically down-hill*), light simply serves to overcome the activation energy and the reaction is said to be **photocatalytic.**

   There are cases where the reaction that is complementary to the target reaction defines the energetics of the overall process (i.e., makes it photosynthetic or photocatalytic). For example,

in the recovery of metal ions M^{n+} from a spent solution by irradiation of a semiconductor (e.g., TiO_2), as in the present experiment, there are two possible cases:

Case a:

$$M^{n+} + A \xrightarrow{h\nu, TiO_2} M^0 + \text{Products of } A$$
$$\Delta G^0 > 0$$

Case b:

$$M^{n+} + B \xrightarrow{h\nu, TiO_2} M^0 + \text{Products of } B$$
$$\Delta G^0 < 0$$

These two cases are exemplified below for M = Cu (see Rajeshwar, 1995 and Reiche, 1979):

a) $2Cu^{2+} + 2H_2O \xrightarrow{h\nu, TiO_2} 2Cu^0 + O_2 + 4H^+$

b) $Cu^{2+} + 2CH_3COO^- \xrightarrow{h\nu, TiO_2} Cu^0$
$$+ C_2H_6 + 2CO_2$$

Calculate ΔG^0 for each process (per mole of Cu) and label them as photosynthetic or photocatalytic.

Note: Take
$E^0_{Cu^{2+}/Cu} = -0.34\,\text{V}, E^0_{O_2/H_2O} = 1.23\,\text{V},$ and
$E^0_{CO_2/CH_3COO^-} = -0.40\,\text{V}.$

Additional Related Projects

- Quantify the amount of Cu(II) recovered after the optional re-dissolution step, by filtering the photocatalyst as described in the experiment and analyzing the resulting solution (for example, by atomic absorption).
- Quantify the amount of CO_2 produced by mass spectrometry or by gravimetry as $BaCO_3$.
- Select other organic molecules with different standard potentials and discuss the role of the semiconductor band locations.
- Analyze the separate effect of adsorption by designing experiments in such a way that this effect can be distinguished from the other effects discussed in the experiment above (see Bumpus, 1999).
- Try other metal ions for recovery as a function of their standard potential.

---

*Answer in this book's website at www.springer.com

- Use light of different wavelengths to observe any possible effects on the process.
- Test other semiconductors with different band gaps.
- Produce hydrogen photocatalytically from a sulfide/sulfite solution irradiated by sunlight (see Koca, 2002).

- Demonstrate a reduction reaction by the photocatalytic reduction of carcinogenic yellow Cr(VI) to the much less dangerous green Cr(III) at UV-illuminated TiO_2 (see Lin, 1993 and Praire, 1993). Use citric acid for the counter (i.e., complementary) reaction at low pH (e.g., pH 2, 0.7 M citric acid, 0.1 weight % of TiO_2, 50 mg/L Cr(VI)).

Name_____Section_____Date_____

Instructor_____Partner_____

LABORATORY REPORT SHEET—EXPERIMENT 16

a) Qualitative procedure

Write down and interpret your observations for each test tube.

Tube #	Physical observations	Interpretation
1		
2		
3		
4		
5		
6		

b) Qualitative procedure

Absorbance before irradiation, A_0: _____

Table of results:

Tube #	Absorbance (after irradiation and filtration), A_t	A_t/A_0
1		
2		
3		
4		
5		
6		

Plot these experimental A_t/A_0 data points. Establish the removal rate equation and interpret it.

POSTLABORATORY PROBLEMS AND QUESTIONS

*1. Upon UV illumination and in the absence of air, a photocatalytic phenomenon occurs in a suspension containing TiO_2 + dissolved copper (II) acetate, $(CuC_4H_6O_4)$ which simultaneously yields the following products (see Reiche, 1979):

a) A metallic deposit (A)
b) A gas (B) that produces a white precipitate upon bubbling in a saturated lime solution (CaO)
c) A gas (C), which upon elemental analysis is shown to be a saturated hydrocarbon. When subject to partial oxidation, this gas generates an acid that contains the same anion present in the initial suspension.

With this information, write:

i) The formulas for A, B, and C
ii) The main (**balanced**) reaction that occurs at:

   I. The semiconductor site acting cathodically:

$$A^{x+} + xe^- = A$$

   II. The semiconductor site acting anodically:

$$xAc^- = xB + C + xe^-$$

   (Note: Ac^- is the acetate anion)
   III. The global reaction

$$A^{x+} + xAc^- = A + xB + C$$

*2. Cyanide ions (CN^-) are very toxic. They can be photocatalytically transformed at the surface of TiO_2 semiconducting particles into the much less problematic isocyanate ions (OCN^-). This oxidation reaction in a *basic medium* requires two holes (h^+) for every cyanide ion to produce one isocyanate ion. On the other hand, the complementary reaction (reduction) that occurs in a different region of the semiconducting particle requires a given number of electrons to reduce the dioxygen present in the solution, thus producing one molecule of hydrogen peroxide.

With this information,

a) Write the balanced oxidation and reduction reactions, and the global reaction (without electrons or holes in it).
b) Would you expect a pH change during the global process? Explain.

Notes:

  i. $TiO_2 + UV = TiO_2 + h^+ + e^-$.
  ii. The recombination of an electron and a hole brings about their "neutralization" in the sense that this process does not yield any substance but only energy.
  iii. A hole can be understood as a "virtual species" in such a way that $A + h^+ = A^+$. On the other hand, $A + e^- = A^-$.

3. Calculate the theoretical number of moles of CO_2 produced from the oxidation of the organic species used in the experiment (assuming a 100% yield).

———————————
* Answer in this book's webpage at www.springer.com

Student Comments and Suggestions

Literature References

Bumpus, J. A.; Tricker, J.; Andrzejewski, K.; Rhoads, H.; Tatarko, M. "Remediation of Water Contaminated with an Azo Dye: An Undergraduate Laboratory Experiment Utilizing an Inexpensive Photocatalytic Reactor," *J. Chem. Educ.* **1999**, *76*, 1680–1683.

Chen, X.; Halasz, S. M.; Giles, E. C.; Mankus, J. V.; Johnson, J. C.; Burda, C. "A Simple Parallel Photochemical Reactor for Photodecomposition Studies," *J. Chem. Educ.* **2006**, *83*, 265–267.

Colon, G.; Hidalgo, M. C.; Navio, J. A. "Photocatalytic Deactivation of Commercial TiO_2 Samples During Simultaneous Photoreduction of Cr(VI) and

Photooxidation of Salicylic Acid," *J. Photochem. Photobiol. A: Chemistry* **2001**, *138 (1)*, 79–85.

Foster, N. S.; Noble, R. D.; Koval, C. A. "Reversible Photoreductive Deposition and Oxidative Dissolution of Copper Ions in Titanium Dioxide Aqueous Suspensions," *Environ. Sci. Technol.* **1993**, *27*, 350–356.

Herrera-Melian, J.A.; Doña-Rodriguez, J. M.; Rendon, E. T.; Soler Vila, A.; Brunet Quetglas, M.; Azcarate, A.; Pascual Pariente, L. "Solar Photocatalytic Destruction of *p*-Nitrophenol: A Pedagogical Use of Lab Wastes," *J. Chem. Educ.* **2001**, *78*, 775–777.

Huang, M.; Tso, E.; Dayte, A. K.; Prairie, M. R.; Stange, B. M. "Removal of Silver in Photographic Processing Waste by TiO$_2$-Based Photocatalysis," *Environ. Sci. Technol.* **1996**, *30*, 3084–3088.

Ibanez, J. G. "Redox Chemistry and the Aquatic Environment: Examples and Microscale Experiments," *Chem. Educ. Int. (IUPAC)* **2005**, *6*, 1–7.

Ibanez, J. G.; Mena-Brito, R.; Fregoso-Infante, A. "Laboratory Experiments on the Electrochemical Remediation of the Environment. Part 8. Microscale Photocatalysis," *J. Chem. Educ.* **2005**, *82*, 1549–1551.

Ibanez, J. G.; Mena-Brito, Rodrigo; Fregoso-Infante, Arturo; Seesing, M.; Tausch, Michael W. "Photoredoxreaktionen mit Titandioxid als Microscale Experimente," *Prax. Naturwiss Chem. Sch.* (Germany, in German) **2005**, *54(3)*, 22–24.

Koca, A.; Sahin, M. "Photocatalytic Hydrogen Production by Direct Sun Light from Sulfide/Sulfite Solution," *Int. J. Hydr. Ener.* **2002**, *27*, 363–367.

Layman, P. L. "Titanium Dioxide Makes Fast Turnaround, Heads for Supply Crunch," *Chem. Eng. News*, Jan. 2, 1995, 12–14.

Legrini, O.; Oliveros, E.; Braun, A. M. "Photochemical Processes for Water Treatment," *Chem. Rev.* **1993**, *93*, 671–698.

Lin, W.-Y.; Rajeshwar, K. "Photocatalytic Removal of Nickel from Aqueous Solutions Using Ultraviolet-Irradiated TiO$_2$," *J. Electrochem. Soc.* **1997**, *144*, 2751–2756.

Lin, W.-Y.; Wei, Ch.; Rajeshwar, K. "Photocatalytic Reduction and Immobilization of Hexavalent Chromium at Titanium Dioxide in Aqueous Basic Media," *J. Electrochem. Soc.* **1993**, *140*, 2477–2482.

Prairie, M. R.; Evans, L. R.; Stange, B. M.; Martinez, S. L. "An Investigation of TiO$_2$ Photocatalysis for the Treatment of Water Contaminated with Metals and Organic Chemicals," *Environ. Sci. Technol.* **1993**, *27*, 1776–1782.

Rajeshwar, K.; Ibanez, J. G. "Electrochemical Aspects of Photocatalysis: Application to Detoxification and Disinfection Scenarios," *J. Chem. Educ.* **1995**, *72*, 1044–1049.

Reiche, H.; Dunn, W. W.; Bard, A. J. "Heterogeneous Photocatalytic and Photosynthetic Deposition of Copper on TiO$_2$ and WO$_3$ Powders," *J. Phys. Chem.* **1979**, *83*, 2248–2251.

Wilkins, F. W.; Blake, D. M. "Use Solar Energy to Drive Chemical Processes," *Chem. Eng. Progr.* **1994**, June, 41–49.

Wilson, E. "Titanium Dioxide Catalysts Break Down Pollutants," *Chem. Eng. News*, Jan. 15, 1996, 23–24.

Experiment 17
Chemical Mineralization of Pollutants Through the Fenton Reaction

Reference Chapter: 10

Objectives

After performing this experiment, the student shall be able to:

- Understand the concept of Advanced Oxidation Processes.
- Produce Fenton's reagent.
- Mineralize an organic sample.
- Observe or monitor the decomposition of a surrogate organic pollutant.
- Compare the effects of different variables on the mineralization of an organic substance (i.e., the presence of UV light, and the nature of the transition metal ion used).

Introduction

As discussed in Chapter 10, pollutant remediation methods are quite varied. Each one has its own merits and challenges. For example, one of the most popular remediation methods, incineration, consists of the elevation of temperature of organic waste in air until combustion occurs. Unfortunately, this is not feasible for many on-site applications, can produce harmful gaseous side-products, requires the generation and handling of high temperatures, and gives rise to non-favorable public reception. As another example, chlorination (bleaching) is used for the discoloration of unwanted colored materials but it may also produce toxic by-products that require further treatment.

These problems can be overcome in many cases through the use of hydrogen peroxide, which is ca-pable of self-decomposing to produce a powerful oxidizer intermediate: the hydroxyl radical, $^\bullet OH$ (see: Advanced Oxidation Processes, AOP in Section 10.1).

As discussed in Section 10.1 of the companion book, hydrogen peroxide is *paradoxically* used both as an oxidizing and as a reducing agent, depending on the redox potential of the pollutant to be treated. H_2O_2 can also yield $^\bullet OH$ radicals in the presence of UV light, although it weakly absorbs solar radiation and thus radical formation by this process is slow. The reaction between Fe^{2+} and H_2O_2 (Fenton's reaction) is given by

$$Fe^{2+} + H_2O_2 = Fe^{3+} + OH^- + {}^\bullet OH \qquad (1)$$

Another photoassisted process involves the generation of H_2O_2 in aqueous solutions at the surface of an illuminated semiconductor (e.g., ZnO), which can in turn form Fenton's reagent with artificially added Fe^{2+}. Interestingly, H_2O_2 can also be generated naturally by active species (mainly radicals) resulting from the photosensitization of humic acids.

The following experiments demonstrate how a surrogate pollutant or hazardous waste (e.g., the organic dye rhodamine B) can be mineralized (i.e., converted into innocuous species like CO_2 and H_2O) upon oxidation in the presence of H_2O_2 under different conditions. We selected this dye due to its resistance to degradation by UV light. The mixture of $Fe^{2+} + H_2O_2$ (*Fenton's reagent*) is highly oxidizing, and so is H_2O_2 when used in conjunction with UV light. In summary, students will be able to compare (either by naked eye observations or by instrumental measurements) the effect of the following substances or conditions: Fe^{2+} alone, H_2O_2 alone,

$H_2O_2 + Mn^{2+}$, $H_2O_2 + Fe^{2+}$ (Fenton's reagent), sunlight, and H_2O_2 + sunlight (see Ibanez, 2003). Mn^{2+} was selected for one of the experiments due to its ability to decompose H_2O_2.

Experimental Procedure

Estimated time to complete the experiment: 2 h (either the instrumental or the non-instrumental options).

Materials	Reagents
1 1-mL graduated pipet	$Ba(OH)_2$
1 2-mL graduated pipet	$Ca(OH)_2$
6 3-mL screw-capped conical bottom	0.02 M H_2SO_4
vials with rubber septa	0.003% rhodamine-B
2 2-mL microburets	D. I. water
2 Beral pipets	$FeSO_4 \cdot 7H_2O$
7 glass capillary tubes	30% H_2O_2
1 test tube (5 mm I.D., 5 cm tall)	$MnCl_2 \cdot 4H_2O$
1 UV-lamp	

With the seven glass capillary tubes make U-shaped capillaries with the aid of a flame from a match or a cigarette lighter (alternatively, plastic capillary tubing may be used). Puncture a tiny hole in the middle of each septum of the conical bottom vials as to allow a capillary tube to go tightly through it. **Caution: Glass capillary tubes are very easy to break. Do not apply any pressure to them. Protect your hands with thick gloves or cloth while doing these insertions and wear eye protection.** Each vial is now connected through one of these U-capillaries to a small test tube (e.g., 5 mm I.D., 5 cm tall) containing a scrubbing solution of $Ba(OH)_2$ (see below) as shown in Figure 1. $Ba(OH)_2$ may be replaced by CaO or $Ca(OH)_2$ if desired, although the results are less indisputable.

Label each vial as A, B, C, D, E, F, and G. Obtain 10 mL of a 0.02 M H_2SO_4 solution. Place 1 mL of this solution and 1 mL of distilled or deionized water in each vial. (Note: In vial D add only 2 mL of water, no acid). Add 10 drops of a dilute (for example, 0.003%) rhodamine-B solution (i.e., 4-[[(4-Dimethylamino)phenyl]-azo]benzenesulfonic acid sodium salt) with a Pasteur pipet to each vial. **Caution: Rhodamine B may be one or more of the following: highly acutely toxic, cholinesterase inhibitor, known/probable carcinogen, known groundwater pollutant or known reproductive or developmental toxicant. Handle with care. The residues containing rhodamine B must be given to the instructor for appropriate handling.**

For qualitative observations, the test tubes that will be used to collect any CO_2 produced during the reactions are now halfway filled with a saturated $Ba(OH)_2$ solution (this yields insoluble white

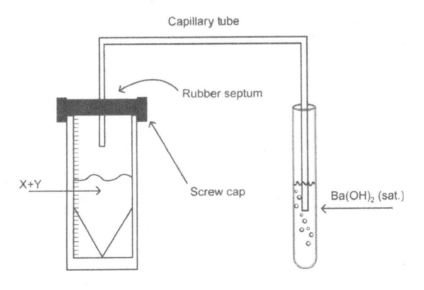

FIGURE 1. Experimental set-up. X + Y represent the reagents specified in each step in the text. (Adapted from Ibanez, 2003).

$BaCO_3$). **Caution: Many barium salts are toxic and must not be ingested nor their dust inhaled**. Connect these tubes to the vials by using the capillary tubes as shown in Figure 1. Close the test tubes loosely with parafilm and use immediately as directed below to prevent any significant absorption of atmospheric CO_2. A compromise must be achieved here, because closing the test tubes too tightly will not allow the gases from the reaction vial to bubble into the scrubbing tubes (with the concomitant pressure build-up and the possibility of the septum blowing-out), whereas closing the test tubes too loosely may allow atmospheric CO_2 to enter the solution and produce misleading results. Then, test for the following effects. Note that vial A (*blank*) only contains sulfuric acid and rhodamine solutions.

1. *The effect of Fe^{2+} alone—Vial B*
   Add 15–20 mg of $FeSO_4 \cdot 7H_2O$ to vial B. Close the vial immediately. Observe, record and interpret any changes in the system after 30 min. **Caution: do not handle the toxic and possibly mutagenic Fe^{2+} with your bare hands. Use appropriate gloves. Do not breathe Fe^{2+}–containing dust.**

2. *The effect of H_2O_2 alone—Vial C*
   Prepare 6–12% H_2O_2 by dilution from a commercial concentrated solution (30%). Add 10 drops to vial C. Close the vial immediately. Observe, record and interpret any changes in the system after 30 min. **Caution: do not handle H_2O_2 with your bare hands, because it is an oxidizer that may attack your skin. Use appropriate gloves.**

3. *The effect of the addition of H_2O_2 and transition metal ions—Vials D and E*
   a. The effect of Mn^{2+}.
   Some transition metal ions (e.g., Mn^{2+}) are known to catalyze the decomposition of H_2O_2. Add 10 drops of 6–12% H_2O_2 and 60 mg of $MnCl_2 \cdot 4H_2O$ to vial D. Sulfuric acid solution was not added to this vial because under such acidic conditions the decomposition of H_2O_2 by Mn^{2+} is not easily observable. Immediately cap the vial and connect the capillary tube (inserted through the septum) to the scrubbing vial. Observe, record and interpret any changes in the system after 30 min.
   b. Production of Fenton's reagent.
   Repeat the above procedure by adding 10 drops of 6–12% H_2O_2 and 15–20 mg of $FeSO_4 \cdot 7H_2O$

to vial E (instead of the Mn salt used above). Immediately cap the vial and connect the U-shaped capillary tube (inserted through the septum) to the scrubbing vial. Observe, record and interpret any changes in the system after 30 min. **Caution: •OH radicals are such powerful oxidizers that various scientists are hypothesizing their possible linkage to some forms of cancer. For this reason, it is advised that students do not breathe nor allow contact with the reaction mixtures/products of steps 3b and 5, and in general, of Fenton's reaction (that is why we strongly recommend to perform these experiments in a closed system). Owing to the short lifetimes of these radicals (in the sub-second range), and to the formation of scavengers of •OH radicals in the scrubbing solution (e.g., carbonate ions), most likely there is no danger associated with such radicals after a few seconds.**

4. *The effect of sunlight—Vial F*
   Add 10 drops of water to vial F to have the same total liquid volume as in the other vials, and expose it to sunlight for 30 min (connected to its scrubbing vial). Observe, record, and interpret any changes in the system after 30 min. Even though Pyrex glass absorbs a large amount of the UV rays contained in the sunlight that reaches the Earth, a small amount of UV light goes through it.

5. *The effect of H_2O_2 + sunlight—Vial G*
   Repeat the procedure in step 4, but this time adding 10 drops of 6–12% H_2O_2 to reaction vial G before exposing the sample to the sunlight for 30 min. Observe, record, and interpret any changes in the system. Note: A UV lamp may also be used for this experiment and for the previous one. Changes are faster and clearer than with direct sunlight exposure, although the hazard of using UV light should be considered. In this case, cover the experimental set up with a cardboard box and a black polyethylene bag. **Caution: UV rays are highly energetic and may cause cancer. Avoid exposure of any parts of your body to UV radiation.**

A more quantitative experiment can be performed by taking the absorption spectrum of the initial, unattacked sample and comparing it to those obtained after each of the above procedures.

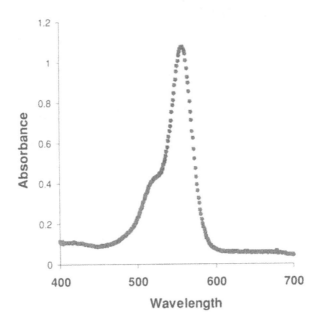

FIGURE 2. Absorption spectrum of the blank sample (rhodamine B). (Reproduced from The Chemical Educator **2003**, *8*, 47–50).

Note: Filtration of the product mixture with a very fine filter medium (e.g., 0.45 μm) is required before taking the absorption spectra if there were a precip- itate in a vial. Note that the maximum absorption wavelength of rhodamine B is 558 nm, as shown in Figure 2.

Name_____Section_____Date_____

Instructor_____Partner_____

PRELABORATORY REPORT SHEET—EXPERIMENT 17

Objectives

Flow sheet of procedure

Waste containment and recycling procedure

PRELABORATORY QUESTIONS AND PROBLEMS

*1. Since the oxidation state of oxygen in H_2O_2 is -1, oxidizers produce O_2 from it, whereas reducers typically produce H_2O or OH^-. Using a clever approach based on oxygen *isotope labeling*, it has been shown that all the O_2 produced during **oxidation** of aqueous H_2O_2 comes from the peroxide bond and not from water. Thus, the O–O bond is *not cleaved* by oxidation.

On the contrary, **reduction** of H_2O_2 by Fe^{2+} (as in Fenton's reaction) adds an extra electron that produces the H_2O_2 radical, which then decomposes into $OH^- + {}^\bullet OH$ (see Baird, 1997). This mechanism *implies cleavage* of the O–O bond, promoted by the extra electron that enters a σ antibonding orbital.

a) Assuming that the bond order in the O–O (i.e., peroxide) bond $= 1$, what is the bond order of the O–O bond in $H_2O_2{}^{-\bullet}$? (Recall from General Chemistry that the bond order, BO is calculated by subtracting the number of electrons that are located in an antibonding orbital in a bond from those in a bonding orbital; the result is then divided by two).

b) Did you obtain the same bond order for both species? What does your finding imply?

*2. When a small (catalytic) amount of Fe^{3+} ions is introduced into an electrochemical cell, it can be reduced to Fe^{2+} at the cathode (reaction 1). Simultaneously, H_2O is oxidized to O_2 at the anode (reaction 2). This O_2 then diffuses to the cathode, where it gets reduced to produce H_2O_2 (reaction 3). When Fe^{2+} and H_2O_2 diffuse away from the electrodes and meet in the solution, Fenton's reaction occurs and ${}^\bullet OH$ radicals are produced (reaction 4). This process is called *electro-Fenton* (see Oturan, 2001).

a) Write and balance each one of the four reactions described above.

b) Using Hess law, combine these four reactions so as to yield a net reaction (reaction 5) of the form:

$$\tfrac{1}{2}A_{(g)} + B_{(l)} \xrightarrow{\text{electrical energy}} 2C_{(aq)}$$

where A is an oxidizing gas, B is a highly

stable polar liquid, and C is an extremely powerful and unstable oxidizer.

c) Why is Fe^{3+} not present in the net reaction (reaction 5)? Explain.

d) What percentage of the O_2 consumed by the system is produced by the anodic reaction (reaction 2)?

3. Fenton reaction can occur in natural waters due to the presence of Fe^{2+} [formed by the photolysis of Fe(III)L complexes], and the presence of H_2O_2 (formed by the photolysis of humic acids and other substances). Write appropriate equations to depict this phenomenon.

Additional Related Projects

• Find in the literature other UV-resistant dyes and test them instead of rhodamine B.

• Quantify the amount of CO_2 produced, by mass spectrometry or gravimetry for example.

• Use UV light of different wavelengths to observe any possible effects on the overall process described in the present experiment.

• Oxidize a small amount (e.g., a few drops) of an aromatic species with Fenton's reagent and do a qualitative test for the presence of the corresponding phenolic derivative **(Caution: many aromatics are highly toxic).** (See Greenberg, 1998).

• Use an Fe(III)L complex (e.g., Fe(III)EDTA, see Experiment 8) to produce Fe(II) by photoreduction. Add H_2O_2 and use the mixture in the same fashion as the normal Fenton's reagent, as described in the procedure above.

• Perform Fenton's reaction on a halogenated aromatic hydrocarbon (e.g., C_2Cl_4) and test for the presence of Cl^- and CO_2 in the products. (See Leung, 1992).

• Repeat the original experiment done by Fenton by adding a drop of an Fe^{2+} solution and a drop of a H_2O_2 solution to a basic solution of tartaric acid (see Fenton, 1894). He reported observing an intense violet color. Could this color be due to the presence of ferrate ions? How can this be tested? Search the literature as needed to see if such a hypothesis is plausible, and if possible, find an analytical tool to test for it. (See Ibanez, 2004).

• Try recently proposed alternative systems that circumvent some of the inherent limitations in the Fenton system (e.g., the requirement of a low

---

working pH, the narrow pH range for solubility of the iron species, and the easiness of oxidation of Fe^{2+} by ambient dioxygen). For exam- ple, try a $Cu(II)/H_2O_2$/organic acid system, or a Co/H_2O_2/ascorbic acid system. (See Verma, 2003 and 2004).

Name_____Section_____Date_____

Instructor_____Partner_____

LABORATORY REPORT SHEET—EXPERIMENT 17

1. Fill-out the following table with your results in the qualitative experiments. Interpret your observations.
See for example the results from Vial D.

Qualitative results

Effect / Reaction vial	Major discoloration, yes or no	Bubbling, yes or no	BaCO$_3$ precipitate, yes or no
A/ Blank			
B/ Fe^{2+} alone			
C/ H$_2$O$_2$ alone			
D/ H$_2$O$_2$ + Mn^{2+}	NO	YES	NO
E/ H$_2$O$_2$ + Fe^{2+}			
F/ UV alone			
G/ H$_2$O$_2$ + UV			

Interpret each one of your results:

Vial A: _____

Vial B: _____

Vial C: _____

Vial D: _____

Vial E: _____

Vial F: _____

Vial G: _____

2. Insert your absorbance vs wavelength spectra for all the samples. Estimate the percentage of the absorbance reduction observed in each sample at the maximum-absorbance wavelength (fill-in the following table). Interpret your observations.

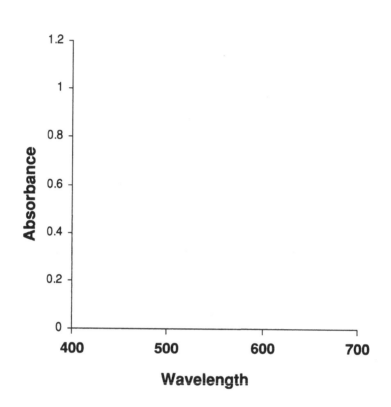

Sample in Vial	Absorbance	Absorbance decrease, %
A		
B		
C		
D		
E		
F		
G		

POSTLABORATORY PROBLEMS AND QUESTIONS

*1. The H_2O_2 necessary for Fenton's reaction can be produced by the photocatalytic generation of holes (h^+, see Section 10.1) that react with water in circumneutral media (reaction a, see below). In alkaline media, however, holes react with OH^- to produce $^\bullet OH$ (reaction b). (See Sanchez, 1996).

a) Write the balanced reaction a in the form:
   $A + B = C + D$
b) Write the balanced reaction b in the form:
   $X + Y = Z$

*2. During the degradation of percholoroethylene, C_2Cl_4 (hereby labeled as A) by Fenton's reagent, the rate limiting step is the *elementary* reaction:

$$A + {}^\bullet OH \xrightarrow{k_1} \text{oxidation products}$$

A plot of ln [A] vs t gave a straight line of slope $-k_2$ (see Leung, 1992). With this information:

a) Show that the concentration of $^\bullet OH$ radicals remains essentially constant throughout the experiment.
b) Write the rate expression as a pseudo-first order expression.

*3. The $^\bullet OH$ radicals produced during *advanced oxidation processes, AOP* (see Chapter 10) can either react with a target pollutant (e.g., A) or—undesirably—with scavengers (e.g., B) as follows:

$$\text{A (target)} + {}^\bullet OH \nearrow \text{C} \quad (1)$$
$$\searrow \text{P} \quad (2)$$

$$\text{B (scavenger)} + {}^\bullet OH \to D \quad (3)$$

where C, P and D are different products. The yield of the first reaction for the production of C after a time t of continuous $^\bullet OH$ production is:

$$\eta_{C,t} = \frac{n_C}{n_{\bullet OH/A}} \quad (4)$$

where n_C is the number of moles of C produced by reaction 1, and $n_{\bullet OH/A}$ is the number of moles of $^\bullet OH$ that reacted with A during t. In the same fashion, the fraction of the total amount of $^\bullet OH$ that

reacted with A during this time is

$$f_{\bullet OH/A} = \frac{n_{\bullet OH/A}}{n_{\bullet OH/(A+B)}} \quad (5)$$

where $n_{\bullet OH/(A+B)}$ is the total number of moles $^\bullet OH$ produced (assuming that all of them reacted with either A or B). (See Derbalah, 2004).

With this information, and assuming i) that the analytical quantification of A, B, and C is possible, ii) that the initial number of moles of A (i.e., $n_{A,0}$) and B (i.e., $n_{B,0}$) are known, and iii) that A and B react only with $^\bullet OH$,

a) Write $\eta_{C,t}$ as a function of n_C and $n_{A,t}$ (where $n_{A,t}$ and $n_{B,t}$ represent the number of moles of A or B, respectively, present at a time t).
b) Write $f_{\bullet OH/A}$ as a function of $n_{A,t}$ and $n_{B,t}$.
c) Write an equation to show that the total number of $^\bullet OH$ produced (i.e., $n_{\bullet OH/(A+B)}$) can be calculated from $f_{\bullet OH/A}$, n_C and $\eta_{C,t}$.

4. Calculate the theoretical number of moles of CO_2 produced during your experiment (assuming a 100% oxidation yield).

---

* Answer in this book's webpage at www.springer.com

Student Comments and Suggestions

Literature References

Baird, C. "Free Radical Reactions in Aqueous Solutions: Examples from Advanced Oxidation Processes for Wastewater and from the Chemistry in Airborne Water Droplets," *J. Chem. Educ.* **1997**, 74, 817–819.

Derbalah, A. S.; Nakatani, N.; Sakugawa, H. "Photocatalytic Removal of Fenitrothion in Pure and Natural Waters by Photo-Fenton Reaction," *Chemosph.* **2004**, 57 (7), 635–644.

Eilbeck, W. J.; Mattock, G. *Chemical Processes in Waste Water Treatment*; Ellis Horwood, Ltd.: Chichister, England, 1987. Chapter 3.

Faust, B. C. "A Review of the Photochemical Redox Reactions of Iron (III) Species in Atmospheric, Oceanic, and Surface Waters: Influences on Geochemical Cycles and Oxidant Formation," Chapter 1 in: Helz, G. R.; Zepp, R. G.; Crosby, D. G., Eds. *Aquatic and Surface Photochemistry*; Lewis Publishers: Boca Raton, FL, 1994.

Fenton, H. J. H. "Oxidation of Tartaric Acid in Presence of Iron," *Chem. Soc. J. Trans.* **1894**, *65*, 899–910.

Greenberg, R. S.; Andrews, T.; Kakarla, P. K. C.; Watts, R. J. "In-Situ Fenton-Like Oxidation of Volatile Organics: Laboratory, Pilot, and Full-Scale Demonstrations," *Remediation* **1998**, *8*, 29–42.

Hong, A.; Zappi, M. E.; Kuo, C. H.; Hill, D. "Modeling Kinetics of Illuminated and Dark Advanced Oxidation Processes," *J. Environ. Eng.* **1996**, *122*, 58–61.

Horikoshi, S.; Hidaka, H.; Serpone, N. "Environmental Remediation by an Integrated Microwave/UV-Illumination Method. 1. Microwave-Assisted Degradation of Rhodamine-B Dye in Aqueous TiO_2 Dispersions," *Environ. Sci. Technol.* **2002**, *36*, 1357–1366.

Ibanez, Jorge G. "Redox Chemistry and the Aquatic Environment: Examples and Microscale Experiments," *Chem. Educ. Int. (IUPAC)* **2005**, *6*, 1–17.

Ibanez, J. G.; Hernandez-Esparza, M.; Valdovinos-Rodriguez, I.; Lozano-Cusi, M.; de Pablos-Miranda, A. "Microscale Environmental Chemistry. Part 2. Effect of Hydrogen Peroxide in the Presence of Iron (II) (Fenton Reagent) and Other Conditions Upon an Organic Pollutant," *Chem. Educator* **2003**, *8*, 47–50.

Ibanez, J. G.; Tellez-Giron, M.; Alvarez, D.; Garcia-Pintor, E. "Laboratory Experiments on the Electrochemical Remediation of the Environment. Part 6. Microscale Production of Ferrate," *J. Chem. Educ.* **2004**, *81*, 251–254.

Leung, S. W.; Watts, R. J.; Miller, G. C. "Degradation of Perchloroethylene by Fenton's Reagent: Speciation and Pathway," *J. Environ. Qual.* **1992**, *21*, 377–381.

Rajeshwar, K.; Ibanez, J. G. *Environmental Electrochemistry: Fundamentals and Applications in Pollution Abatement*; Academic Press: San Diego, CA, 1997. Chapter 5.

Oturan, M. A.; Oturan, N.; Lahitte, C.; Trevin, S. "Production of Hydroxyl Radicals by Electrochemically Assisted Fenton's Reagent. Application to the Mineralization of an Organic Micropollutant, Pentachlorophenol," *J. Electroanal. Chem.* **2001**, *507*, 96–102.

Sanchez, L.; Peral, J.; Domenech, X. "Degradation of 2,4-Dichlorophenoxyacetic Acid by *In Situ* Photogenerated Fenton Reagent," *Electrochim. Acta* **1996**, *41*, 1981–1985.

Verma, P.; Baldrian, P.; Gabriel, J.; Trnka, T.; Nerud, F. "Copper-Ligand Complex for the Decolorization of Synthetic Dyes," *Chemosph.* **2004**, *57*, 1207–1211.

Verma, P.; Baldrian, P.; Nerud, F. "Decolorization of Structurally Different Synthetic Dyes Using Cobalt (II)/Ascorbic Acid/Hydrogen Peroxide System," *Chemosph.* **2003**, *50*, 975–979.

Experiment 18
Production and Analysis of Chloramines

Reference Chapter: 10

Objectives

After performing this experiment, the student shall be able to:

- Produce the three chloramines.
- Understand the distribution of chloramine species a function of pH.
- Analyze the chloramines spectrophotometrically.

Introduction

A thorough discussion of chloramines, their uses, advantages, and problems, is presented in Section 10.1 of the companion book. Hypochlorous acid can react stepwise with ammonia so as to produce the three chloramines (i.e., mono-, di- and trichloramine):

$$NH_{3(g)} + HClO \rightarrow NH_2Cl + H_2O_{(l)} \quad (1)$$

$$NH_2Cl + HClO \rightarrow NHCl_2 + H_2O_{(l)} \quad (2)$$

$$NHCl_2 + HClO \rightarrow NCl_3 + H_2O_{(l)} \quad (3)$$

The predominance diagram as a function of pH for the three chloramines is given in Figure 1, and their species distribution diagram is given in Figure 2.

Chloramine Analysis

Chloramines can be analyzed by titrimetry, colorimetry (with N,N-diethyl-p-phenylenediamine, DPD), voltammetry, UV spectrophotometry, am-

perometry, mass spectrometry, HPLC, and other methods.

In the present experiment students will produce the three chloramines and characterize them qualitatively by UV spectroscopy, taking advantage of the simplicity of this analytical method.

Experimental Procedure

Estimated time to complete the experiment: 1.5 h.

Materials	Reagents
3 10-mL volumetric flasks	NH₄Cl
1 1-mL graduated pipet	10 M NaOH
2 Beral pipets	3 M HCl
3 spectrophotometer cuvettes	6% NaOCl
3 5-mL screw-capped conical bottom vials with rubber septa	0.01 M H₂SO₄
3 glass capillary tubes	
1 10-mL syringe	
1 spectrophotometer	
1 pH meter	

Caution: All the reactions for the production of chloramines must be done under a fume hood (see the hazards listed below). Note: A "portable" fume hood, made by packing a 20-mL syringe with activated charcoal, could be appropriate for this microscale production (i.e., the *Obendrauf trap*, see Corral, 2005, although its efficiency with chloramines remains to be analyzed).

In a 10-mL volumetric flask prepare an approximately 0.1 M NH₄Cl solution and adjust the pH to 10–11 with a 10 M NaOH solution. This can be done, for example, by dissolving 53.5 mg of NH₄Cl in

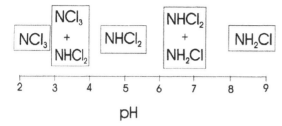

FIGURE 1. Approximate chloramine distribution with pH (for a chlorine $(1+)$ to nitrogen $(3-)$ ratio, R $<$ 1). If R $>$ 1 the predominance zones are displaced towards higher pH values; for example, NCl_3 under such conditions can in principle be stable even at pH $=$ 8. (Data taken from Colin, 1987).

5 mL of water, adding the NaOH solution as needed and using the necessary amount of water to reach the 10-mL mark. With a pipet or a syringe, transfer 3-mL aliquots of this solution into three spectrophotometer cuvettes. Add to the first cuvette five drops of deionized (D.I.) water; add to second cuvette three drops of DI water and two drops of 3 M HCl, and add to the third cuvette five drops of 3 M HCl. Stir. Measure the pH of each one (these should be in the approximate ranges 10–11, 6–7, and 2–3, respectively).

Prepare chlorine gas in a 5-mL screw-capped vial, equipped with a septum and a U-glass capillary exiting through the septum so as to convey gases from

the vial into the spectrophotometric cuvette. See Figure 3. Chlorine gas preparation can be done for example by adding 1 mL of 3 M HCl to the vial, followed by injection with a 10-mL syringe of one mL of commercial hypochlorite solution (typically 6%) plus 9 mL of air (see Mattson, 2003). Air is used here as the carrier so as to direct the chlorine gas that forms in the vial into the first spectrophotometric cuvette containing NH_4Cl solution. Allow this process of pushing the chlorine gas into the reaction mixture to last for 3 min. Measure the pH of the resulting solution and immediately take the corresponding absorbance spectrum in the region 200–400 nm. Repeat this procedure with the other two spectrophotometer cuvettes.

The solutions remaining in the vials and in the spectrophotometer cuvettes can be neutralized under a fume hood with HCl or NaOH as needed, and discarded according to local regulations.

Hazards (see Delalu, 2001; Tanen, 1999; Pepi, 2003):

- The reaction of monochloramine with excess NH_3 in alkaline media produces hydrazine, a known carcinogen. [Fortunately, at the working pH (about 10) the amount of NH_3 in excess of NH_4^+ is small and it may react with chlorine gas to produce the chloramine].

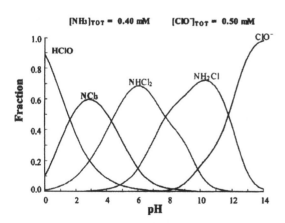

FIGURE 2. Species distribution diagram for the chloramines as a function of pH (at the arbitrary conditions: $[NH_3]_{tot} = 0.40$ mM and $[ClO^-]_{tot} = 0.50$ mM). (Reproduced from *The Chemical Educator*, see Ibanez, 2006, with permission).

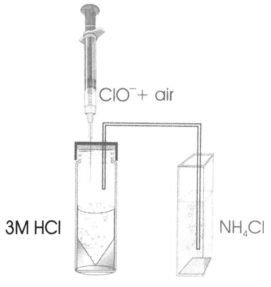

FIGURE 3. Experimental set-up. (Adapted from Ibanez, 2006).

- Accidental mixing of household cleaners containing hypochlorite and ammonia may result in severe lung injury.
- NCl_3 is the most unstable chloramine in the pure form and it is highly explosive, a strong irritant, and a lacrimator.

- Concentrated NaOH is corrosive to human tissue (eye protection is especially important).
- Avoid mixing hypochlorite with any acid since this may result in the production of hazardous gaseous chlorine.

Name_____Section_____Date_____

Instructor_____Partner_____

PRELABORATORY REPORT SHEET—EXPERIMENT 18

Objectives

Flow sheet of procedure

Waste containment procedure and recycling procedure

PRELABORATORY QUESTIONS AND PROBLEMS

*1. Electrochemical studies of the three chloramines show that the mono- is reduced first in an electrochemical cell (i.e., requires the least negative or more positive potential), then the di- and lastly, the trichloramine. The generalized reaction is:

$$NH_{3-n}Cl_n + xH^+ + ye^- = zCl^- + NH_4^+$$

Fill the following table with the stoichiometric coefficients (i.e., n, x, y, z) involved in the three reactions, corresponding to the reduction of one mole of each one of the chloramines.

Chloramine	n	x	y	z
Monochloramine				
Dichloramine				
Trichloramine				

*2. If one mole of chloramine underwent *comproportionation* (i.e., the contrary of *disproportionation*, see Example 2.11 in the companion book) with 1 mole of ammonia in a basic medium to yield 1 mole of hydrazine, N_2H_4 as the sole nitrogen-containing product plus one mole of hydrochloric acid, which one of the three chloramines was the starting material?

*3. Oxidation of ammonia by chlorine may produce various compounds such as: NO, N_2O_4, N_2H_4, N_2, N_2O, NO_2^-, NO_3^-, and NH_2OH.

a) Arrange such compounds according to the oxidation state of the nitrogen contained in each one.

b) Fill-in the following table with the requested formulas and the stoichiometric coefficients that correspond to the theoretical oxidation with elemental chlorine of each nitrogenated species (symbolized as X), as in the generalized equation:

$$aNH_3 + bCl_2 + cH_2O = X + yCl^- + zH^+$$

(The case of N_2O is given as an example).

Oxidation state of N in X	Formula of X	a	b	c	y	z	Moles of Cl_2 reduced/mol of ammonia nitrogen
-2							
-1							
0							
$+1$	N_2O	2	4	1	8	8	$4/2 = 2$
$+2$							
$+3$							
$+4$							
$+5$							

Additional Related Projects

- Alternative dichloramine preparations. (Note: The resulting solutions can be analyzed by UV spectrometry as described in the experiment above).
  1. React hypochlorite with ammonia (in 10% excess) in a dilute basic medium at $0°C$ to prepare the monochloramine. Then, add slowly dilute perchloric acid to the resulting solution (to a pH of 3.5–4) to prepare the dichloramine. An excess of acid will produce the trichloramine.
  2. Prepare the monochloramine and then make the resulting solution trickle down a cation-exchange resin bed. As it turns out, there exists a quite acidic medium in the interior of the resin grains and, as a result, the NH_2Cl undergoes dismutation to yield $NHCl_2 + NH_4^+$.
- Trichloramine concentration. Produce the trichloramine as described in the present experiment and then extract it with an organic solvent (for example, try *baby oil*) so as to obtain a more concentrated solution. Run a UV-VIS scan as above and compare it to that obtained with the non-extracted solution prepared earlier (see Pepi, 2003).
- Once the absorbance spectra are taken for each chloramine solution in the present experiment, change the pH of each solution and predict and verify the new major absorbance peaks.

---

* Answer in this book's webpage at www.springer.com

Name_____Section_____Date_____

Instructor_____Partner_____

LABORATORY REPORT SHEET—EXPERIMENT 18

Insert the absorbance spectra obtained for the three solutions. What are the wavelengths at the highest peaks obtained with the solutions from each spectrophotometric cuvette?

Peak of solution 1_____nm. pH = _____

Peak of solution 2_____nm. pH = _____

Peak of solution 3_____nm. pH = _____

POSTLABORATORY PROBLEMS AND QUESTIONS

*1. The formation of monochloramine by the following reaction in the aqueous phase:

$$HClO_{(aq)} + NH_{3(aq)} = NH_2Cl_{(aq)} + H_2O_{(l)}$$

is an *elementary* reaction with a rate constant, $k_2 = 5.1 \times 10^6$ L mol^{-1} s^{-1} (see Vernon, 1980). Because the distribution of species in the HClO/ClO$^-$ ($pK_a = 7.5$) and NH_4^+/NH$_3$ ($pK_a = 9.3$) systems is pH dependent, the overall rate must be pH dependent as well. If the total concentration of free chlorine (I) species (see Section 10.1.1.1) is given by:

$$C_{T,Cl(I)} = [HClO] + [ClO^-]$$

then one can use the definition of f (i.e., the fraction of species in a given form) given in Section 2.1 of the companion book and obtain the fraction of HClO, f_{HClO} as:

$$f_{HClO} = [HClO]/C_{T,Cl(I)}$$
$$= [HClO]/\{[HClO] + [ClO^-]\}$$

In the same fashion, the total concentration of ammonia species, $C_{T,NHx}$ is

$$C_{T,NHx} = [NH_4^+] + [NH_3]$$

and the corresponding ammonia fraction, f_{NH3} is

$$f_{NH3} = [NH_3]/C_{T,NHx}$$
$$= [NH_3]/\{[NH_4^+] + [NH_3]\}$$

You may assume that $C_{T,Cl(I)}$ and $C_{T,NHx}$ are large enough as to remain constant in the first moments of the reaction, and that kinetic data are taken there.

With such data (and without performing any mathematical calculations), answer the following questions (a–d).

a) What is the reaction order? Write the rate equation of monochloramine formation as a function of k_2, $C_{T,Cl(I)}$, $C_{T,NHx}$, f_{HClO}, and f_{NH3}

b) Sketch the species distribution diagram (defined in Section 2.1) in the HClO/ClO$^-$ system

c) Sketch the species distribution diagram in the NH_4^+/NH$_3$ system

d) On the basis of your responses to a, b, and c, select the graph that best describes the pH-dependent behavior of the rate of this reaction (see the graph options below).

_____
* Answer in this book's webpage at www.springer.com

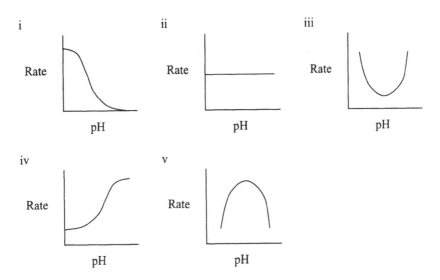

i   ii   iii

Rate   pH

Rate   pH

Rate   pH

iv   v

Rate   pH

Rate   pH

*2. Repeat steps *b* and *c* from the previous problem, but this time perform the numerical calculations and plot each *f* vs pH diagram.

*3. With the data from problems 1 and 2, calculate and plot the rate values as a function of pH in the pH range of 0–14. For simplicity, assume that $C_{T,Cl(I)} = C_{T,NHx} = 1 \times 10^{-4}$ M. Compare the shape of this plot to your selection in question *1d*.

4. Calculate the theoretical concentration of chloramine produced in each of your vials, assuming a 100% yield and that the limiting reagent is NH_4^+.

Student, Comments and Suggestions

Literature References

Apolinari, L.; Kowalski, T. "UV-Spectrophotometric Studies of some Properties of Chloramines", *Prace Naukowe—Inst. Inzyn. Ochrony Srodowiska Polit. Wroclaws* (Poland, in Polish) **1978**, *40*, 89–100.

Colin, C.; Brunetto, M.; Rosset, R. "Les Chloramines en Solution: Preparations, Equilibres, Analyse," *Analusis* (France, in French) **1987**, *15(6)*, 265–274.

Corral, L. E.; Rodríguez-Ibáñez, O.; Fernández-Sánchez, L. "Activated Carbon Traps to Avoid Pollutants in the Chemistry Laboratories," *Educ. Quim.* (Mexico, in Spanish) **2005**, *16(3)*, 114–117.

Delalu, H.; Peyrot, L.; Duriche, C.; Elomar, F.; Elkhatib, M. "Synthesis of Enriched Solutions of Chloramine Starting from Hypochlorite at High Chlorometric Degree," *Chem. Eng. J.* **2001**, *83*, 219–224.

Harrington, G. W.; Noguera, D. R.; Kandou, A. I.; Vanhoven, D. J. "Pilot-scale Evaluation of Nitrification Control Strategies," *J. Am. Wat. Works Assoc.* **2002**, *94 (11)*, 78–89.

Ibanez, J. G.; Balderas-Hernandez, P.; Garcia-Pintor, E.; Espinoza-Marvan, L.; Ruiz-Martin, R. M.; Arrieta, J. J.; Ramírez-Silva, M. T.; Casillas, N. "Microscale Environmental Chemistry. Part 7. Properties, Preparation, and Characterization of Chloramines," *Chem. Educator* **2006**, *11*, 402–405.

Mattson, B.; Anderson, M.; Mattson, S. *Microscale Gas Chemistry*, 3rd ed.; Educational Innovations: Norwalk, CT (USA), 2003.

Maui Department of Water Supply, Wailuku, HI (USA). http://www.mauiwater.org/chloramines.html

Norton, C. D.; LeChevallier, M. W. "Chloramination: Its Effect on Distribution System Water Quality," *J. Am. Wat. Works Assoc.* **1997**, *89 (7)*, 66–77.

Pepi, F.; Ricci, A.; Rosi, M. "Gas-phase Chemistry of $NH_xCl_y^+$ ions. 3. Structure, Stability, and Reactivity of Protonated Trichloramine," *J. Phys. Chem.* **2003**, *107*, 2085–2092.

Shang, C.; Blatchley, E. R. III. "Differentiation and Quantification of Free Chlorine and Inorganic Chloramines in Aqueous Solution by MIMS," *Environ. Sci. Technol.* **1999**, *33*, 2218–2223.

Stumm, W.; Morgan, J. J. *Aquatic Chemistry: Chemical Equilibria and Rates in Natural Waters;* Wiley Interscience: New York, 1996. Chapter 11.

Tanen, D. A.; Graeme, K. A.; Raschke, R. "Severe Lung Injury after Exposure to Chloramine Gas from Household Cleaners," *New Engl. J. Medic.* **1999**, *341*, 848–849.

Vernon, L. S. *Water Chemistry;* Wiley: New York, 1980. Chapter 7.

Wilczak, A.; Hoover, L. L.; Lai, H. H. "Effects of Treatment Changes on Chloramine Demand and Decay," *J. Am. Wat. Works Assoc.* **2003**, *95 (7)*, 94–106.

Wrona, P. K. "Electrode Processes of Chloramines in Aqueous Solutions," *J. Electroanal. Chem.* **1998**, *453 (1–2)*, 197–204.

Experiment 19
Production and Analysis of Chlorine Dioxide

Reference chapter: 10

Objectives

After performing this experiment, the student shall be able to:

- Produce a relatively uncommon, environmentally friendly, oxidizing and disinfecting gas.
- Visualize an application of Frost diagrams.
- Perform a disproportionation reaction.
- Produce the same substance by different paths (i.e., oxidation, reduction and disproportionation).

Introduction

As discussed in Section 10.1 of the companion book, chlorine gas is a potent and relatively cheap oxidizer and disinfectant. Unfortunately, its reaction products (disinfection by-products, or DBP) can be quite dangerous. A chlorine-based alternative to chlorine itself is chlorine dioxide, ClO_2 (also called chlorine peroxide) that can act as an extremely effective biocide, disinfectant, and oxidizer. It is active against some chlorine-resistant pathogens. Additional advantages include that its oxidizing and disinfecting properties remain essentially constant over a wide pH range (from 4 to 10), and that its DBPs are substantially fewer than those produced by chlorine. A summary of its applications, economic perspectives, molecular geometry and properties is given in Section 10.1.

Chlorine dioxide can be produced in many different ways, mainly through the reduction of Cl(V) or the oxidation of Cl(III). The following equations

in aqueous solution are only indicative of the main reactions:

a) Chemical reduction of Cl(V)

$$ClO_3^- + 1/2 H_2O_{2(1)} + H^+ \rightarrow$$
$$ClO_{2(g)} + 1/2 O_{2(g)} + H_2O_{(1)} \quad (1)$$

$$ClO_3^- + 1/2 CH_3OH_{(1)} + H^+ \rightarrow$$
$$ClO_{2(g)} + \text{other products} \quad (2)$$

$$ClO_3^- + 1/2 H_2SO_4 + 1/2 SO_{2(g)} \rightarrow$$
$$ClO_{2(g)} + HSO_4^- \quad (3)$$

$$1/2 ClO_3^- + 1/2 ClO_2^- + H^+ \xrightarrow{cat}$$
$$ClO_{2(g)} + 1/2 H_2O_{(1)} \quad (4)$$

$$ClO_3^- + Cl^- + 2H^+ \rightarrow$$
$$ClO_{2(g)} + 1/2 Cl_{2(g)} + H_2O_{(1)} \quad (5)$$

b) Chemical oxidation or disproportionation of Cl(III)

$$ClO_2^- + 1/2 Cl_{2(g)} \rightarrow ClO_{2(g)} + Cl^- \quad (6)$$

$$ClO_2^- + 1/2 ClO^- + 1/2 H_2O_{(1)} \xrightarrow{H^+}$$
$$ClO_{2(g)} + OH^- + 1/2 Cl^- \quad (7)$$

$$5/4 HClO_2 \xrightarrow{cat.} ClO_{2(g)} + 1/2 H_2O_{(1)}$$
$$+ 1/2 Cl^- + 1/2 H^+ \quad (8)$$

$$ClO_2^- + xO_{3(g)} \xrightarrow{pH 5-6} ClO_{2(g)} + \text{products} \quad (9)$$

The production of ClO_2 by chlorate reduction or by chlorite oxidation can also be performed electrochemically. Several of the routes described in the reactions above also entail the production of Cl_2 as a by-product, which is undesirable for certain applications; it can be removed by contacting the gas mixture with oxides, hydroxides, and various carbonates of the alkali and alkaline earth metals. In addition, the newer processes (e.g., reduction of chlorate with hydrogen peroxide, reaction 1) are designed to minimize the amount of Cl_2 generated as by-product.

Analytical methods for chlorine dioxide include:

- Spectrophotometry at 360 nm (extinction coefficient = 1150 M^{-1} cm^{-1})
- Titrimetry with iodine
- Polarography
- Amperometry
- Chromatography

In the present experiment we describe three microscale procedures to obtain chlorine dioxide: a) Cl(V) reduction, b) Cl(III) oxidation, and c) Cl(III) disproportionation. The gas produced is analyzed spectrophotometrically either in the gas phase or dissolved in water.

Experimental Procedure

Estimated time to complete the experiment: 1.5 h.

Materials	Reagents
3 10-mL Erlenmeyer flasks	5 M H_2O_2
3 10-mL beakers	30% H_2O_2
3 5-mL syringes	D.I. water
1 universal stand	0.4 M $NaClO_2$
1 three-finger clamp	NaClO
3 2-mL graduated pipets	2.5 M $NaClO_3$
3 Beral pipets	10 M H_2SO_4
1 propipet bulb	0.01 M H_2SO_4
1 spectrophotometer	0.2 M $NaClO_2$
3 spectrophotometer cuvettes	2 M H_2SO_4
1 ice bath	
1 10-mL graduated cylinder	
1 pH meter	

Each one of the three methods described below for the preparation of $ClO_{2(g)}$ utilizes the apparatus shown in Figure 1. The apparatus setup is as follows: First stretch the plastic stem of a 4-mL, narrow stem disposable transfer pipet (Beral pipet) by holding the tip in one hand and the bulb in the other hand and pulling in opposite directions slowly and firmly. This will make a long capillary-like tube (30–45 cm) that allows for easy delivery of the gas. Submerge the tip of the pipet in a small Erlenmeyer flask

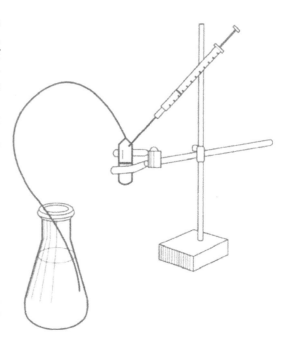

FIGURE 1. Experimental set-up. (Adapted from Ibanez, 2006).

(or preferably, in a spectrophotometer cuvette) filled with distilled or deionized water. The reactants are then introduced sequentially in each experiment, as described below. As chlorine dioxide gas is evolved from the reaction in the bulb, it travels through the pipet stem and into the flask or cell, where it readily dissolves in water. The solubility of chlorine dioxide in water increases with decreasing temperature; therefore, it is helpful to place the receiving flask or cell in an ice bath during the reaction. Injecting several mL of air into the pipet bulb serves to push the product ClO_2 out of the bulb and into the receiving flask or cell. The aqueous solution can then be used for characterization and/or further experiments. Owing to the high solubility of this gas, the analytical test described below can be applied either to the gas or to the collection solution.

Caution: All the preparations described below must be performed under a fume hood, since inhalation of ClO_2 may cause respiratory irritation, pulmonary edema and even death. Do not wear contact lenses during its preparation or use. Whenever production of ClO_2 is to be stopped (for example, in a chemical spill) make the solution alkaline so as to convert it to ClO_4^- and ClO_2^-, thus avoiding further gas release. In case

of contact with skin or eyes, flush immediately and abundantly with lukewarm water. Before voluntary release to the atmosphere, ClO_2 should be reacted or decomposed; appropriate media for ClO_2 removal include a solid soda-lime fine mixture, a basic thiosulfate solution, the Witches brew (i.e., a mixture of sodium hydroxide and potassium salts), etc. Do not allow ClO_2 pressure to build up, since at p >150 mm Hg it will likely decompose explosively! It decomposes thermally and photochemically with expansion.

Preparation method #1: Chlorate reduction by hydrogen peroxide

Caution: Sodium chlorate is a strong, explosive oxidizer. Do not allow contact with organic matter or other oxidizable substances.

In a 10-mL beaker, place 7.5 mL of distilled water, 0.1 mL of a freshly prepared 2.5 M sodium chlorate solution, 0.2 mL of a freshly prepared 5 M hydrogen peroxide solution (this can be prepared, for example, by diluting 5.7 mL of 30% H_2O_2 with D.I. water to a total of 10 mL), and 1.0 mL of a 0.01 M sodium chloride solution. **Caution: Concentrated H_2O_2 is**

harmful to human tissue. Use skin and eye protection.**

Draw approx. 2 mL of this mixture into the Beral pipet through its stem (before stretching it as described above); then, stretch it and puncture a small hole on one of its sides and introduce **very slowly** through it (e.g., with a 5-mL syringe) 1-2 mL of 10 M sulfuric acid to initiate the reaction. See Figure 1. **Caution: Sulfuric acid is corrosive, especially when concentrated; use skin and eye protection. Syringe needles are very sharp, handle with care.** (If possible, unsharpen the needles with sandpaper or a fine metal file to convert them into "steel tubes"). Push the gas product out of the pipet and collect it for analysis in the manner described above. A yellow color (indicative of the production of ClO_2) appears immediately. Pump more air with the syringe by injecting several syringes-full of air into the pipet so as to displace the gas. Alternatively, an aquarium air pump or a compressed inert gas (e.g., nitrogen) can provide the necessary carrier gas.

An aqueous solution of chlorine dioxide has a distinctly yellowish cast, whereas chlorine gas dissolved in water is more green. Because $Cl_{2(g)}$ is often a by-product in the production of $ClO_{2(g)}$, it may be

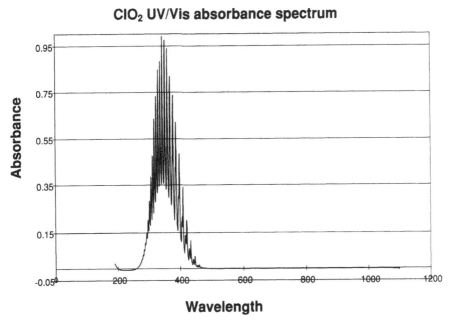

ClO_2 UV/Vis absorbance spectrum

FIGURE 2. Gas-phase chlorine dioxide absorbance. (Adapted from Ibanez, 2006).

difficult to distinguish by eye if the solution is more yellow or green, and therein whether the reaction produced the desired product. UV-Vis spectra may be taken (of either the aqueous solution or the gas) to confirm the presence of ClO_2. A peak for chlorine dioxide gas is observed at 360 nm with distinctive vibrational fine structure (see Figure 2). This electronic spectrum is discussed in detail elsewhere (see Esposito, 1999). It may be helpful to take a scan of chlorine gas to compare it to the chlorine dioxide spectrum.

Preparation method #2: Chlorite oxidation by hypochlorite

Caution: Sodium chlorite is an oxidizing, corrosive material. Combustibles wetted with its solution and subsequently dried are easily ignited and burn vigorously (e.g., paper). It is incompatible with all combustibles, reducing agents (including reactive metals) and acids. It is a severe (corrosive) irritant. In case of spill of the solid material, collect into a clean metal or high-density polyethylene container. Wash away residues with a large amount of water. Do not use rags, sawdust or other combustible absorbents. If contact with the skin occurs, wash with soap and water.

Pour a few mL of a 0.4 M $NaClO_2$ solution into a 10-mL beaker. Neutralize it to pH 7 with 0.01 M H_2SO_4. In a separate container, prepare a working solution of sodium hypochlorite by diluting 0.2 mL of a commercial (6% w/v) solution (bleach) to 10 mL with D.I. water and neutralize to pH 7 with 0.01 M H_2SO_4. Draw approximately 1.5 mL of this dilute bleach solution into a new Beral pipet through its stem (before stretching it as described above); then, stretch it and puncture a small hole on one of its sides and introduce through it (with a syringe) 1.5 mL of the neutralized NaClO solution. Proceed as in the preparation method #1 to displace, collect, and analyze the ClO_2. See the cautionary notes given above.

Preparation method #3: Disproportionation of chlorous acid

Draw approximately 1.5 mL of a 0.2 M $NaClO_2$ solution into a Beral pipet through its stem (before stretching it as described above); then, stretch it and puncture a small hole on one of its sides and introduce through it (with a syringe) approximately 1.5 mL of a 2 M sulfuric acid solution. Proceed as in the preparation method #1 to displace, collect, and analyze the ClO_2. See the cautionary notes given above.

Name_____Section_____Date_____

Instructor_____Partner_____

PRELABORATORY REPORT SHEET—EXPERIMENT 19

Experiment Title: _____

Objectives

Flow sheet of procedure

Waste containment and recycling procedure

PRELABORATORY QUESTIONS AND PROBLEMS

*1. Predict whether ClO_2 is thermodynamically stable or unstable at $pH = 0$ and $pH = 14$ by drawing the Frost diagram of chlorine in its different oxidation states (see how to construct a Frost diagram in Section 2.3 of the companion book).

*2. In the series of reduction reactions for the production of ClO_2 given in the introduction, which one is a *simple* comproportionation?

*3. In the series of reduction reactions for the production of ClO_2 given in the introduction, which one is a *double* comproportionation?

*4. In the series of chemical oxidation or disproportionation reactions of Cl(III) for the production

of ClO_2 given in the introduction, which one a *simple* disproportionation?

*5. In the series of chemical oxidation or disproportionation reactions of Cl(III) for the production of ClO_2 given in the introduction, which ones are *double* disproportionations?

Additional Related Projects

- Produce ClO_2 by electrochemical reduction of chlorate ions or the oxidation of chlorite ions (see Rajeshwar, 1997).
- Analyze the amounts of ClO_2 produced by a given technique when varying key parameters.
- Condense ClO_2 as a red liquid.
- Trap ClO_2 selectively with a glycine solution from a gas stream containing ClO_2 and Cl_2.

---

*Answer in this book's webpage at www.springer.com

Name_____Section_____Date_____

Instructor_____Partner_____

LABORATORY REPORT SHEET—EXPERIMENT 19

Preparation method #1: Chlorate reduction by hydrogen peroxide

Number of moles of chlorate used   _____

Number of moles of hydrogen peroxide used   _____

Number of moles of sulfuric acid used   _____

Wavelength used for the UV-Vis analysis of ClO_2 _____

Absorbance   _____

Calculated concentration of ClO_2 _____

Preparation method #2: Chlorite oxidation by hypochlorite

Number of moles of $NaClO_2$ used   _____

Number of moles of hypochlorite used   _____

Absorbance   _____

Calculated concentration of ClO_2 _____

Preparation method #3: Disproportionation of chlorous acid

Number of moles of $NaClO_2$ used   _____

Number of moles of sulfuric acid used   _____

Absorbance   _____

Calculated concentration of ClO_2 _____

POSTLABORATORY PROBLEMS AND QUESTIONS

*1. We have produced ClO_2 in a simultaneous fashion at the anode and cathode of an electrochemical cell. This process in which both electrodic reactions yield the same product is a *convergent paired electrosynthesis* (see Paddon, 2006). The cathodic reaction consists of the reduction of chlorate ions, whereas the anodic reaction is the oxidation of chlorite ions. The cation in the catholyte as well as in the anolyte is Na^+, which remains inert during the process. Assuming that all the current passing through the electrodes is employed for the production of ClO_2,

a) Write the balanced anodic reaction (neutral medium)
b) Write the balanced cathodic reaction (acidic medium)
c) Write the global (net) reaction (acidic medium). Note that this reaction is a comproportionation.

*2. Chlorine dioxide can be produced by the *disproportionation* of chlorous acid as follows:

$$a\,HClO_2 \rightarrow 4X + 2H_2O + Y^- + H^+$$

Balance this equation.

*3. Chlorine dioxide can be produced by the *double comproportionation* of two different acids (both contain Cl) as follows:

$$b\,HClO_3 + Z \rightarrow c\,ClO_2 + d\,Cl_2 + H_2O$$

(Z does not contain oxygen).

Balance this equation.

Student Comments and Suggestions

---

*Answer in this book's website at www.springer.com

Literature References

ERCO Worldwide. http://clo2.com/index.html

Esposito, A.P.; Stedl, T.; Jonsson, H.; Reid, P. J.; Peterson, K. A. "Absorption and Resonance Raman Study of the $^2B_1(X) - {}^2A_2(A)$ Transition of Chlorine Dioxide in the Gas Phase," *J. Phys. Chem. A*, **1999**, *103*, 1748–1757.

Green, T. J.; Islam, M.; Canosa-Mas, C.; Marston, G.; Wayne, R. P. "Higher Oxides of Chlorine: Absorption Cross-sections of Cl_2O_6 and Cl_2O_4, the Decomposition of Cl_2O_6, and the Reactions of OClO with O and O_3," *J. Photochem. Photobiol. A: Chemistry* **2004**, *162*, 353–370.

Ibanez, J. G.; Navarro-Monsivais, C.; Terrazas-Moreno, S.; Mena-Brito, R.; Pedraza-Segura, L. Mattson, B.; Anderson, M. P.; Fujita, J.; Hoette, T. "Microscale Environmental Chemistry, Part 5. Production of ClO_2, an Environmentally-Friendly Oxidizer and Disinfectant," *Chem. Educ.* **2006**, *11*, 174–177.

McKetta, J. *Inorganic Chemicals Handbook*. Vol. 2, "Chlorine Dioxide"; Marcel Dekker: New York, 1993. pp. 779–810.

Oikawa, K.; Hayashi, Y. "Method for Preparing Stabilized Aqueous Chlorine Dioxide Solution," U.S. Patent 5,165,910 (Nov. 24, 1992).

Paddon, C. A.; Atobe, M.; Fuchigami, T.; He, P.; Watts, P.; Haswell, S. J.; Pritchard, G. J.; Bull, S. D.; Marken, F. "Towards Paired and Coupled Electrode Reactions for Clean Organic Microreactor Electrosyntheses," *J. Appl. Electrochem.* **2006**, *36*, 617–634.

Rajeshwar, K.; Ibanez, J.G. *Environmental Electrochemistry: Fundamentals and Applications in Pollution Abatement*: Academic Press: San Diego, 1997. Chapter 7.

Roozdar, H. "Method and Compositions for the Production of Chlorine Dioxide," U.S. Patent 5,651,996. (Jul. 29, 1997).

Rosenblatt, A.; Rosenblatt, D. H.; Feldman, D.; Knapp, J. E.; Battisti, D.; Morsi, B. "Method and Apparatus for Chlorine Dioxide Manufacture," U.S. Patent 5,234,678 (Aug. 10, 1993).

Sokol, J. C. "Process for Continuously Producing Chlorine Dioxide," U.S. Patent 5,380,517 (Jan. 10, 1995).

Svenson, D. R.; Kadla, J. F.; Chang, H-M.; Jameel, H. "Effect of pH on the Inorganic Species Involved in a Chlorine Dioxide Reaction System," *Ind. Eng. Chem. Res.* **2002**, *41*, 5927–5933.

Yin, G.; Ni, Y. "Mechanism of the ClO_2 Generation from the H_2O_2-$HClO_3$ Reaction," *Can. J. Chem. Eng.* **2000**, *78*, 827–833.

Experiment 20
Metal Ion Recovery By Cementation

Reference Chapters: 10, 12

Objectives

After performing this experiment, the student shall be able to:

- Remove a metal ion from a solution by using a more active metal.
- Understand the concept of cementation.
- Analyze the rate of removal of an ion from a solution.
- Interpret the physical and chemical phenomena observed during the course of a cementation process.

Introduction

As discussed in Chapter 10, the uncontrolled release of metal ions poses a threat to the environment. An alternative removal scheme is *cementation*, whereby redox activity takes place between a metal substrate, M_1, and the ions of a more noble metal, M_2 (i.e., a metal with a more positive standard potential that needs to be removed from the solution), due to a thermodynamically spontaneous reaction.

In general, a cementation process can be described by the following equation (for simplicity, two divalent metals are considered here):

$$M^{2+}_{(aq)} + N^0_{(s)} = M^0_{(s)} + N^{2+}_{(aq)} \quad (1)$$

In this way, N is the *sacrificial* metal, and M is the *cemented* metal (that is, the one to be removed). Besides the spontaneous nature of the process, other advantages include its relative simplicity, ease of

control, and the possible recovery of valuable metals. Challenges to this process involve the possible presence of chelating agents, mass transfer limitations, and the introduction of new species into the medium. Some examples of applications of cementation processes include:

- removal of polluting metal ions from aqueous effluents
- purification steps in metallurgical operations
- recovery of spent metals from different chemical operations

The present experiment is an example of this last application. However, in environmental cleanup applications one has to analyze the possible impact of the new ions added to the aqueous system.

Experimental Procedure

Estimated time to complete the experiment: 2 h

Materials	Reagents
1 (or more, according to the instructor) 3-mL spectrophotometer cuvettes	0.02, 0.04, 0.06, 0.08 and 0.1 M CuSO$_4$
1 spectrophotometer	1 Zn pellet

This experiment comprises two steps. 1) The cementation of Cu^{2+} ($E^0_{Cu^{2+}/Cu} = 0.34$ V) by Zn ($E^0_{Zn^{2+}/Zn} = -0.76$ V) is accomplished in a small-volume (about 3 mL) semimicro spectrophotometer cuvette (path length = 1 cm), and the concentration

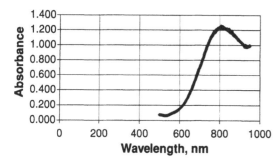

FIGURE 1. Absorbance spectrum of $CuSO_4$ in solution. (Adapted from Ibanez, 2007).

of Cu^{2+} is monitored *in situ* at a fixed, predetermined wavelength. 2) A calibration plot is obtained with known solutions; then, absorbance versus time data points are acquired and transformed into a concentration versus time curve from which removal kinetic data can be analyzed.

a) Calibration curve

Prepare a series of at least five different concentrations of $CuSO_4$ solutions (for example, 0.02, 0.04, 0.06, 0.08, and 0.1 M). Trace the absorption vs wavelength curve with a scanning spectrophotometer from 400 to 900 nm for each solution. The peak in the vicinity of 800–810 nm is used for the quantification of Cu^{2+} in this experiment. See Figure 1. A fixed-wave spectrophotometer (or colorimeter)

can also be used at 810 nm. (Although more precise $Cu(II)$ determinations exist in the literature, this strategy dramatically simplifies the experimental procedure). Then, obtain a calibration curve by plotting the absorbance vs concentration points for each concentration.

b) Cementation of Cu^{2+}

Find a Zn pellet of appropriate dimensions such that it cannot fall to the bottom of the spectrophotometer cuvette. Rinse the pellet with acetone or ethanol to remove any grease or other foreign substances from its surface, allow it to dry, and place it in the cuvette as shown in Figure 2.

Next, with a Pasteur or a Beral pipet fill the cuvette with 0.1 M $CuSO_4$ solution as to completely cover the Zn pellet. Insert the cuvette into the spectrophotometer and take an absorbance vs time datum every 5 min. Convert each absorbance point to concentration by using your calibration curve. Allow the experiment to proceed until one half or more of the initial amount of Cu^{2+} is removed. Plot the results on a concentration vs time curve.

At the end of the experiment, separate the Cu-covered Zn pellet and place it in a labeled container. The instructor may choose to recycle it for future experiments. Collect all the solutions containing $Cu(II)$. The instructor may choose to recover the Cu from these solutions (e.g., by electrolysis).

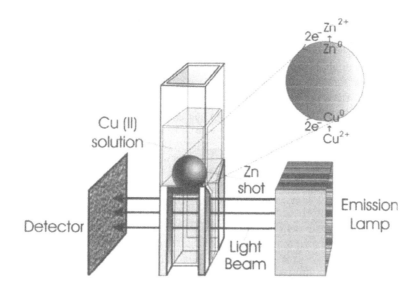

FIGURE 2. Experimental set-up. (Adapted from Ibanez, 2006).

Name_____Section_____Date_____

Instructor_____Partner_____

PRELABORATORY REPORT SHEET—EXPERIMENT 20

Objectives

Flow sheet of procedure

Waste containment and recycling procedure

PRELABORATORY QUESTIONS AND PROBLEMS

1. Give three examples of real cases where the cementation technique is utilized for the removal and/or recovery of metal ions.

2. If a metal ion (M_1^{n+}) is to be cemented and it has a standard potential of $+ 0.50$ V (vs the standard hydrogen electrode, SHE), which of the following metals (M_2) could be used to achieve that operation? Select the standard potential of the metal half reaction that would be useful. a) 0 V, b) $+ 0.50$ V, c) -0.75 V, d) $+ 1.00$ V vs SHE. Explain.

*3. Predict whether the following reactions are spontaneous or not, assuming $T = 25°C$ and that all concentrations are 1 M. Find the necessary data in any table of standard potentials.

a) $2Ag_{(s)} + Ni^{2+}_{(aq)} \rightarrow 2Ag^{+}_{(aq)} + Ni_{(s)}$

b) $Ca_{(s)} + Cd^{2+}_{(aq)} \rightarrow Ca^{2+}_{(aq)} + Cd_{(s)}$

c) $2Br^{-} + Sn^{2+}_{(aq)} \rightarrow Br_{2(1)} + Sn_{(s)}$

Additional Related Project

- As stated in Section 10.1, zero-valent metals (ZVM) can be used not only for the recovery of less-reactive metals, but also for the construction of barriers to the diffusion of other reducible pollutants (e.g., CrO_4^{2-}, NO_3^{-}, toxic halogenated organics, organic ketones, explosives, solvents, dyes) in underground reservoirs and currents. Students can work on projects aimed at the reduction of such pollutants with different ZVM (e.g., Fe, Zn, Al). (See for example the paper by Balko, 2001).

*Answer in this book's webpage at www.springer.com

Name_____Section_____Date_____

Instructor_____Partner_____

LABORATORY REPORT SHEET—EXPERIMENT 20

Observations

1. Color of the initial solution   _____

2. Original color of the solid metal   _____

3. Color of the final solution   _____

4. Insert your calibration curve here, including the equation obtained and the correlation coefficient:

5. Calculate the number of moles of Cu(II) removed   _____moles

6. Insert your concentration vs time plot here

7. Literature reports typically show first order kinetics for the removal of ions by cementation. Test for this hypothesis as applied to your system by plotting the corresponding equation from Section 3.2 of the companion book.

POSTLABORATORY PROBLEMS AND QUESTIONS

*1. Write a general equation for the removal of a metal ion, $M_{(aq)}^{x+}$ from an aqueous solution by cementation on a solid metal, $N_{(s)}^0$ that becomes $N_{(aq)}^{y+}$.

*2. Comment on any possible deviations from linearity in your kinetic plot.

*3. During the removal of Cu^{2+} ions by cementation on elemental Fe, the following reactions occur:

$$Cu_{(aq)}^{2+} + 2e^- = Cu_{(s)} \tag{1}$$

$$Fe_{(s)} = Fe_{(aq)}^{2+} + 2e^- \tag{2}$$

If the system is allowed to reach thermodynamic equilibrium, then the concentration of Cu^{2+} ions will be greatly reduced, whereas the concentration of Fe^{2+} will increase.

What will be the ratio of the concentrations at equilibrium, $Q = [Fe^{2+}]/[Cu^{2+}]$?

(Note: Assume that the concentrations of the ions are approximately the same as their activities).

---

* Answer in this book's webpage at www.springer.com

Student Comments and Suggestions

Literature References

Balko, B. A.; Tratnyek, P. G. "A Discovery-Based Experiment Illustrating How Iron Metal Is Used to Remediate Contaminated Groundwater," *J. Chem. Educ.* **2001**, *78*, 1661–1664.

Bólser, D. G. "Method for Recovering Metals from Solutions," U.S. Patent 5,679, 259. Oct. 21, 1997.

Chang, C. M.; Gu, H.; O'Keefe, T. J. "Review of the Galvanic Stripping Process for Use in Treating Oxidized Metal Wastes," Hazardous Substance Research Center, and the Waste Management Education and Research Consortium (HSRC/WERC) Joint Conf. on the Environ. 1996, Albuquerque, N. M., USA, May 21–23, 1996. Abstract No. 41.

Hebermann, W.; Haag, A.; Kochanek, W. "Removal of Nobler Metal Ions than Iron from Process and Waste Waters," U.S. Patent 5,169,538. Dec. 8, 1992.

Hsu, Y. J.; Kim, M. J.; Tran, T. "Electrochemical Study on Copper Cementation from Cyanide Liquors Using Zinc," *Electrochim. Acta* **1998**, *44*, 1617–1625.

Ibanez, J. G.; Lopez-Mejia, E.; Echevarria-Eugui, J. A. "Microscale Environmental Chemistry. Part 8. Removal of Metal Ions by Cementation," *Chem. Educator* **2007** (in press).

Ku, Y.; Chen, C-H. "Removal of Chelated Copper from Wastewaters by Iron Cementation," *Ind. Eng. Chem. Res.* **1992**, *31*, 1111–1115.

Markhloufi, L.; Saidani, B.; Cachet, C.; Wiart, R. "Cementation of Ni^{2+} Ions from Acidic Sulfate Solutions onto a Rotating Zinc Disk," *Electrochim. Acta* **1998**, *43*, 3159–3164.

Mishra, K. G.; Paramguru, R. K. "Some Electrochemical Studies on Cementation of Copper onto Zinc from Sulfate Bath," *J. Electrochem. Soc.* **2000**, *147*, 3302–3310.

Rajeshwar, K.; Ibanez, J. G. *Environmental Electrochemistry: Fundamentals and Applications in Pollution Abatement*; Academic Press: San Diego, CA, 1997. Chapter 5.

Shokes, T. E.; Moller, G. "Removal of Dissolved Heavy Metals from Acid Rock Drainage Using Iron Metal," *Environ. Sci. Technol.* **1999**, *33*, 282–287.

Wragg, A. A.; Bravo de Nahui, F. N. "Copper Recovery from Aqueous Waste Streams Using Packed and Fluidised Bed Cementation; Influence of Process Parameters," in *Electrochemical Engineering and the Environment 92*, ICHEME Symp. Ser. No. 127: Rugby, England, 1992. pp. 141–152.

Experiment 21
Green Chemistry: The Recovery and Reuse of Sulfur Dioxide (Obendrauf's Method)

Reference Chapters: 10, 12

Objectives

After performing this experiment, the student shall be able to:

- Capture toxic SO_2 in an aqueous solution.
- Produce $CaSO_4$ as a recovery product.
- Produce different gases using plastic syringes (i.e., SO_2, CO_2, O_2).

Introduction

The combustion of hydrocarbons for the production of thermal energy usually involves the production of undesired "chimney gases." In theory these gases should only contain $CO_2 + H_2O$, but in reality they contain important amounts of polluting gases (mainly NO_x, SO_x, CO). A common strategy for the removal and recovery of the SO_2 consists in reacting it with a base [typically $Ca(OH)_2$] to produce a non-toxic substance ($CaSO_3$). However, this salt is not of industrial interest. A more important substance from the industrial standpoint is gypsum, $CaSO_4$, that is a useful material for construction purposes.

In Chapter 12 we defined Green Chemistry as *the design of chemical products and processes that reduce or eliminate the use and generation of hazardous substances*. Because the present experiment involves an environmentally benign chemical synthesis using a waste substance, it falls within the scope of Green Chemistry.

The present desulfurization modeling experiment (called Obendrauf's method) is based on the reaction between SO_2 and $CaCO_3$ to produce

$Ca(HSO_3)_2$, which is then oxidized by dioxygen to the sparingly soluble $CaSO_4$. In fact, $CaSO_3$ is such a good dioxygen scavenger that it has been under study as a means of preventing acid formation during pyrite oxidation (discussed and exemplified in Experiment 11 on *acid mine drainage*). (See Hao, 2000). No hazardous wastes are produced.

Experimental Procedure

Estimated time to complete the experiment: 1.5 h

Materials	Reagents
1 13 mm – test tube	$Ca(OH)_2$
3 10-mL syringes	$NaHCO_3$
1 universal stand	HCl
1 three-finger clamp	3% H_2O_2
1 Beral pipet	$NaHSO_3$
4 rubber stoppers	2 M H_2SO_4
1 ice bath	0.001 M $KMnO_4$
1 microscope slide	0.01 M $Na_2S_2O_3$
1 microscope (optional)	$CaCl_2$

The overall experimental sequence can be seen in Figure 1.

Place 3 mL of a saturated $Ca(OH)_2$ solution in a small test tube. Now produce CO_2 in a 10-mL syringe (by any standard method; see for example Obendrauf, 1996, 2004 and Mattson, 2003) and bubble 5 mL of it through the above solution. A precipitate then forms (i.e., $CaCO_3$). Next, produce SO_2 in a syringe (by any standard method, see for example Obendrauf, 1996, 2004 and Mattson, 2003) and bubble 5 mL of it through this solution until

Desulfurization and gypsum formation

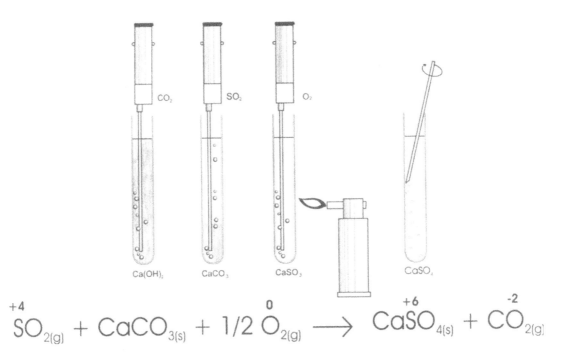

$$\overset{+4}{SO}_{2(g)} + CaCO_{3(s)} + 1/2 \overset{0}{O}_{2(g)} \longrightarrow \overset{+6}{CaSO}_{4(s)} + \overset{-2}{CO}_{2(g)}$$

FIGURE 1. Overall experiment. (Adapted from Obendrauf, 1996).

the precipitate dissolves (try to deliver very small bubbles).

The $CaCO_3$ precipitate will dissolve due to its reaction with SO_2 to form the soluble salt $Ca(HSO_3)_2$. Then, produce O_2 in a syringe (by any standard method, see for example Obendrauf, 1996, 2004 and Mattson, 2003), warm the test tube containing the $Ca(HSO_3)_2$ to 45–50° C for 5–10 min, and bubble 5 mL of O_2 through this solution. The sparingly soluble $CaSO_4$ forms. However, in order to observe the precipitate one often needs to cool down the tube and its contents to room temperature (or even below with an ice bath or with cooling spray).

Scratch the inner wall of the tube with a glass rod so as to produce nucleation sites for the $CaSO_4$. Tiny crystals appear that can be observed through a microscope by placing a few drops of the resulting solution on a microscope slide.

If no precipitate were observed, one can take advantage of LeChatelier's principle by using the *common ion effect* (for example, add a few drops of a saturated $CaCl_2$ solution or—better—a couple of small $CaCl_2$ crystals). This addition of extra Ca^{2+} will push the equilibrium toward the products (i.e., where the sulfate is) and $CaSO_4$ will form a milky precipitate.

Name_____Section_____Date_____

Instructor_____Partner_____

PRELABORATORY REPORT SHEET—EXPERIMENT 21

Objectives

Procedure Flow sheet of procedure

Waste containment and recycling procedure

PRELABORATORY QUESTIONS AND PROBLEMS

1. Fill out the following table.

Compound	Solubility at 25°C (g compound/100 g H_2O)
$CaSO_3$	
$CaSO_4$	
$CaCl_2$	

2. How much SO_2 can be theoretically produced from the burning of 1 L of unleaded regular gasoline in an automobile engine (without catalytic conversion)? (Students may be requested to calculate this for one or more gasoline vendors—Exxon, Shell, Conoco, etc.)

3. Find at least three industrial uses for gypsum ($CaSO_4$).

4. Calculate and plot a species distribution diagram for sulfite in the pH range 0–14 (see Chapter 2 of the companion book). From here, which sulfur-containing species undergoes oxidation in the present experiment?

Additional Related Projects

- Produce gypsum by a similar process to that in the present experiment, but using Na_2CO_3 instead of $CaCO_3$, and adding $CaCl_2$ to convert the final sodium salt to gypsum. (See Voeste, 1979).
- Add a Fe(II) or Fe(III) salt in the oxidation step of the present experiment and evaluate its possible catalytic effect. (See Karatza, 2005 and Trzepierczynska, 1991).
- Regenerate SO_2 from gypsum by reacting the later with H_2S. (See Foecking, 1971).
- Use MgO or $Mg(OH)_2$ instead of the corresponding calcium compounds for the present experiment. Transform the $MgSO_4$ thus produced into gypsum by reacting it with $CaCO_3$. (See Ueno, 1979).

Name_____Section_____Date_____

Instructor_____Partner_____

LABORATORY REPORT SHEET—EXPERIMENT 21

a) Write the balanced reactions for the following steps that occur in the present experiment:

• Production of CO_2

• Production of $CaCO_3$

• Production of SO_2

• Production of $CaSO_3$

• Production of O_2

• Production of $CaSO_4$

b) From these balanced reactions, calculate the number of moles of each substance of interest produced in each step.

POSTLABORATORY PROBLEMS AND QUESTIONS

* 1. Fossil fuel combustion gases contain SO_2 that can be used to produce sodium and calcium sulfates. Replace the italic letters in the following reaction scheme with plausible chemical formulas. (See Satrio, 2002.)

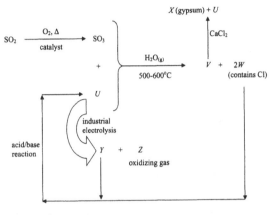

*2. Sulfur can be recovered from gypsum in the form of elemental S and H_2SO_4 through the following (simplified) reaction sequence (see Campbell, 1971):

Reaction a:
$CaSO_4$ is heated in the presence of a mixture of reducing gases (i.e., H_2 and CO) to produce five binary compounds: $U + V + W + X + 4Y$ for every two moles of $CaSO_4$. The last four binary compounds produced are oxides (basic, acidic, acidic, and amphoteric, respectively).

Reaction b:
U reacts with W and Y to produce $H_2S + T$.

Reaction c:
Heating T produces $V + W$.

Reaction d:
H_2S comproportionates with X to give elemental sulfur $+ Y$.

Reaction e:
X is oxidized catalytically in air to produce Z, which reacts with Y to produce H_2SO_4.

---

*Answer in this book's webpage at www.springer.com

A) Find the chemical formulae of T, U, V, W, X, Y, Z.

B) Complete and balance the following reaction scheme:

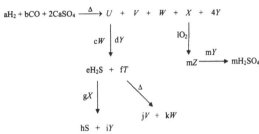

*3. Since the oxidation step in the warm sulfite solution of the present experiment could be incomplete, how can one prove that the precipitate obtained in the experiment is actually $CaSO_4$ and not hydrated $CaSO_3$? (Note that both are sparingly soluble). Suggest a physical and a chemical test to prove this.

Student Comments and Suggestions

Literature References

Angevine; P. A.; Bengtsson, S.; Koudijs, G. P. "Method for Oxidation of Flue Gas Desulfurization Absorbent and the Product Produced Thereby," U.S. Patent 4,544,542. Oct. 1, 1985.

Campbell, R. E.; Fisher, E. E. "Sulfur Recovery from Calcium Sulfate," U.S. Patent 3,607,068. Sept. 21, 1971.

Delplancq, E.; Casti, P.; Vanderschuren, J. "Kinetics of Oxidation of Calcium Sulfite Slurries in Aerated Stirred Tank Reactors," *Chem. Eng. Res. Des.* **1992**, *70(A3)*, 291–295.

Foecking, R. M.; Austin, R. D. "Recovery of Sulfur from Gypsum," U.S. Patent 3,607,036. Sept. 21, 1971.

Goodwin, R. W. "Oxidation of Flue Gas Desulfurization Waste and the Effect on Treatment Modes," *J. Air Poll. Contr. Assoc.* **1978**, *28 (1)*, 35–39.

Hao, Y.; Dick, W. A. "Potential Inhibition of Acid Formation in Pyritic Environments Using Calcium Sulfite Byproduct," *Environ. Sci. Technol.* **2000**, *34*, 2288–2292.

Harrison, D. P. "Regeneration of Sulfided Sorbents and Direct Production of Elemental Sulfur," NATO ASI Series, Series G: *Ecol. Sci.* **1998**, *42* (Desulfurization of Hot Coal Gas), 331–364.

Heinz, H.; Heinrich, I.; Heinrich, G.; Heribert, D. "Semi-Dry Flue Gas Desulphurisation Process Having CaSO₄ as End Product with Simultaneous Separation of NOx," German Pat. DE3318407, Nov. 22, 1984.

Iwatsuki, M.; Ishizaka, H.; Yoshikawa, H.; Oda, N. "Wet Stack Gas Desulfurizing Apparatus and Oxidative Air Supplying Apparatus," Jap. Pat. 2003112011, Apr. 15, 2003.

Karatza, D.; Prisciandaro, M.; Lancia, A.; Musmarra, Dico. "Kinetic and Reaction Mechanisms of Calcium Bisulfite Catalytic Oxidation," *Chem. Eng. Sci.* **2005**, *60*, 1497–1502.

Lancia, A.; Musmarra, Dico; Pepe, F.; Prisciandaro, M. "Model of Oxygen Absorption into Calcium Sulfite Solutions," *Chem. Eng. J.* **1997**, *66 (2)*, 123–129.

Mandelik, B. G.; Pierson, C. U. "New Source for Sulfur," *Chem. Eng. Progr.* **1968**, *64 (11)*, 75–81.

Mattson, B.; Anderson, M.; Mattson, S. *Microscale Gas Chemistry*, 3rd ed.; Educational Innovations: Norwalk, CT (USA), 2003.

Mo, J.; Wu, Z. "Experimental Study on Co-Precipitation of Calcium Sulfite and Calcium Sulfate," American Chemical Society Nat. Meet., Div. Environ. Chem. San Diego, CA. March 13–17, 2005. pp. 318–323.

Obendrauf, V. "Kein Pech mit dem Schwefel—Low-Cost Versuche mit SO₂ und H₂S," *Chem. Sch.* (Austria, in German) **1996**, *11*, 12–15.

Obendrauf, V. "Experimente mit Gasen im Minimasstab," *Chemie in unserer Zeit* (Austria, in German) **2004**, *30 (3)* 118–125.

Rathi; R. J.; Benson, L. B. "Flue Gas Desulfurization with Oxidation of Calcium Sulfite in FGD Discharges," U.S. Patent 4,976,936. Dec. 11, 1990.

Robinson, M. C.; Kirk, D. W.; Hummel, R. L. "Process to Produce Hydrogen and/or Hemihydrate Calcium Sulphate From Calcium Sulphite," U.S. Patent 5,458,744. Oct. 17, 1995.

Satrio, J. A. B.; Jagtap, S. B.; Wheelock, T. D. "Utilization of Sulfur Oxides for the Production of Sodium Sulfate," *Ind. Eng. Chem. Res.* **2002**, *41*, 3540–3547.

Trzepierczynska, I.; Gostomczyk, M. A. "Oxidation of Calcium Sulfite Coming from Flue Gas Desulfurization in the Presence of an Iron Catalyst," *Environ. Protect. Eng.* **1991**, *16 (1)*, 99–110.

Ueno, R.; Kamaya, T. "Desulfurization of Flue Gas," Japanese patent, Jpn. Tokkyo Koho JP 54,035,192. 1979.

Voeste, Th. "Removal of Sulfur Oxides from Combustion Gases," Ger. Offen. DE 78-2820357. 1979.

Zielke, C. W.; Lebowitz, H. E.; Struck, R. T.; Gorin, E. "Sulfur Removal During Combustion of Solid Fuels in a Fluidized Bed of Dolomite," *J. Air Poll. Contr. Assoc.* **1970**, *20 (3)*, 164–169.

Experiment 22
Microorganisms in Soil, Water, and Air

Reference Chapters: 7, 11

Objectives

After performing the experiment, the student shall be able to:

- Isolate bacteria, yeast, and fungi from soil, air, or water samples with the aid of selective media.
- Observe the morphology of the colonies obtained and identify their major physical characteristics.
- Stain when necessary and observe the results in the microscope.

Introduction

Microorganisms have a considerable impact in the environment as well as in our daily lives, as they can gradually modify an ecosystem, cause disease to plants, animals, and humans or they may be used for the production of compounds such as antibiotics, vaccines, enzymes added to detergents, or food such as wine and yoghurt.

Thanks to the genetic manipulation of microorganisms, they can be used to synthesize products foreign to their natural metabolism (e.g., the human hormone insulin, bovine growth hormone, and the factor VIII involved in coagulation). It is important to isolate bacteria, yeasts, and filamentous fungi from their natural habitats, and to be able to purify, characterize and classify them.

An important distinction is their capacity to live under various environmental conditions, sometimes even in extreme conditions of temperature, pH, and salinity that other organisms cannot tolerate.

In this experiment, students will isolate microorganisms from different environments and observe them through an optical microscope. The method used here is the agar plate method using selective media that allows for the growth of specific types of microorganisms [i.e., nutrient agar is used for the growth of bacteria, while potato dextrose agar (*PDA*) is preferred for yeast growth, and Sabourad medium for filamentous fungi].

Although this isolation method has several limitations, it is simple and can give an idea of the diversity of microorganisms in different environments. For example:

- anaerobes in soil fail to grow on the surface of agar exposed to air
- non-symbiotic, nitrogen-fixing organisms grow under these conditions only to a limited extent
- many of the cellulose-digesting forms fail to grow or grow poorly on nutrient agar

Therefore, the observed microorganisms represent only a fraction of the total viable bacterial or yeast population.

Experimental Procedure

Estimated time to complete the experiment: 3 h the first day and 30 min after 24–48 h

When doing microbiological laboratory work, the following rules must be followed:

1. Use a protective garment (e.g., a lab coat)

2. Never eat or drink in the laboratory, and avoid placing objects in your mouth

3. Always wash your hands before leaving the lab

4. If you spill living organisms, cover the spilled material with paper towels and pour a laboratory disinfectant over the towels and the entire contaminated area. Wait 15 min before you clean it up

6. Using a sterile pipet, transfer 0.1 mL of the 10^{-2} dilution to another test tube containing 9.9 mL of the saline solution, and repeat the step in order to have 10^{-4} and 10^{-6} dilutions. Mix thoroughly.

Use an aseptic technique when making serial dilutions and plating. Always use a clean, sterile pipet for all transfers.

Materials	Reagents
20 sterile Petri dishes, 60 × 15 mm	Dextrose potato agar
10 1-mL pipets	Nutrient agar
3 500-mL Erlenmeyer flasks	Sabourad agar
1 mixing plate	Saline isotonic solution (NaCl 0.9%)
1 Bunsen burner	Gram reagents, solutions of:
8 screw cap 13 × 100 mm test tubes	– crystal violet stain
Test tube rack	– Gram's iodine solution
Autoclave	– acetone-alcohol mixture 70:30
2 Incubators (37°C and 30°C)	– safranine stain
Inoculating loop	
Glass slides and cover slips	
Optical microscope	
Immersion oil	
Soil and water samples	
Sterile spatula	
Sterile mortar and pestle	

1. Prepare 100 mL of an isotonic saline solution. Add 9.9 mL to each of six tubes and cap them. Sterilize for 15 minutes at 121°C.

2. Sterilize six 1-mL pipets.

3. Prepare the necessary amount of nutrient agar, of Sabourad agar, and of potato dextrose agar, as indicated by the manufacturers (each Petri dish requires 10 mL of the medium). Mix thoroughly. Heat gently and bring the mixture to boil. Autoclave for 15 min at 15 psi and 121°C. Maintain at 45°C until they are poured in sterile Petri dishes.

Soil sample

4. If necessary, place some soil in a sterile mortar and break up the lumps with a pestle.

5. Weigh 0.1 g of the sample soil. Suspend in the 9.9 mL of saline solution and mix thoroughly. (This means a 1:100 dilution, or 10^{-2}).

7. Label each Petri dish (as 1–7) and add aliquots of the dilutions and media according to the following table:

Dilutions and media	Petri dishes	1	2	3	4	5	6	7
10^{-2}	mL	1	----	----	----	----	1	----
10^{-4}	mL	----	1	----	1	----	----	1
10^{-6}	mL	----	----	1	----	1	----	----
PDA agar	mL	10	10	10				
Nutrient agar	mL				10	10		
Sabourad agar	mL						10	10

8. Slowly, move each Petri dish so that the samples become mixed with the culture medium.

9. Let the agar solidify and incubate (at 37°C for nutrient agar and 30°C for PDA and Sabourad media).

Water sample

10. Measure 0.1 mL of the liquid sample, dilute in 9.9 mL of saline solution and mix thoroughly. (This means a 1:100 dilution, or 10^{-2}).

11. Repeat steps 6–9.

Air sample

12. Prepare Petri dishes with each agar medium.

13. When the agar has solidified, open the dish for 20 min to allow airborne microorganisms contact the medium.

14. Close and incubate, as in step 9.

15. Observe and count the cultures after 24 and 48 h for bacterial growth, and 48–72 h for yeast and fungi growth.

Bacteria

1. Observe the macroscopic characteristics of the colonies of microorganisms that grew on each plate.
2. Prepare separate smears of the unknown bacteria in the following way:
   - Place a drop of tap water at the center of a glass slide.
   - Using the inoculating loop, aseptically remove half a loopful of bacteria from the culture.
   - Mix the large amount of organisms on the loop into the drop of water on the slide.
   - Spread the mixture.
   - Flame the loop to prevent contamination of the worktable, your culture and yourself.
   - Allow the smear to air dry.
   - Heat-fix by passing the slide back and forth through the flame of your Bunsen burner.
3. Perform the Gram stain as follows:
   - Flood the slide with crystal violet and allow to react for 1 min.
   - Handle the slide with slide forceps. Tilt it to an angle of approximately 45° and drain the dye off the slide into a pan or staining sink.
   - Continue to hold the slide at 45° and immediately rinse it thoroughly with a gentle stream of water from the wash bottle.
   - Flush the slide with iodine solution. Allow the iodine to react for 1 minute.
   - Tilt the slide and allow it to drain.
   - Immediately rinse the slide thoroughly with water.
   - With the slide still held at a 45° angle, decolorize it by allowing the acetone–alcohol mixture to run over and off the smear.
   - Rinse immediately with water.
   - Flood the slide with the safranine counterstain. Allow the countestain to react for 1 minute.
   - Rinse the slide thoroughly with water.
   - Carefully blot your stained slide and examine each smear in the microscope under high power and oil immersion objectives.

The Gram stain separates bacteria into two groups, depending on whether the original stain (violet crystal) is retained or lost when the stained smear is treated with an iodine solution and then washed with the acetone–alcohol mixture.

Organisms that retain the stain when washed with alcohol are termed *Gram positive*; those that fail to retain the original stain but take the counterstain (safranine) are called *Gram negative*. The Gram stain is of considerable value in identifying and classifying bacteria.

Yeast and fungi

Yeast and fungi are observed directly under the microscope by placing a loopful of each colony in a drop of water between a slide and a cover slip. Observe the form, size, and number of different colonies under the microscope (at high power and oil immersion objectives).

Autoclave all Petri dishes when the experiment is concluded to kill the isolated microorganisms

---

Name_____Section_____Date_____

Instructor_____Partner_____

PRELABORATORY REPORT SHEET—EXPERIMENT 22

Objectives:

Flow sheet of procedure

Waste containment procedure:

PRELABORATORY QUESTIONS AND PROBLEMS

1. Why are bacteria and fungi classified in different kingdoms?

2. What is the difference between yeasts and filamentous fungi?

3. What are the mean sizes of these microorganisms?

4. What is a selective medium?

5. Describe the shapes of bacteria and yeasts.

Additional Related Projects:

• The dilution factors may have to be changed as the original soil or water sample used will have different microorganism concentrations. There should be fewer than 50 colonies per Petri dish with the highest dilution aliquot.

• Dilutions can be plated by triplicate so that the mean number of colony-forming units (*CFU*)/mL present in the sample can be determined.

Name_____Section_____Date_____

Instructor_____Partner_____

LABORATORY REPORT SHEET—EXPERIMENT 22

SOIL SAMPLE

1. Sketch the macroscopic appearance of the colonies isolated from each medium.

PDA agar	Nutrient agar	Sabourad agar
◯	◯	◯

2. Describe the microscopic appearance of the different isolated microorganisms:

Bacteria	Form	Gram stain
1		
2		
3		
4		
5		
Yeast	**Form**	
1		
2		
3		
4		
5		
Fungi	**Form**	
1		
2		
3		
4		

WATER SAMPLE

1. Sketch the macroscopic appearance of the colonies isolated from each medium.

PDA agar	Nutrient agar	Sabourad agar
◯	◯	◯

2. Describe the microscopic appearance of the different isolated microorganisms:

Bacteria	Form	Gram stain
1		
2		
3		
4		
5		
Yeast	**Form**	
1		
2		
3		
4		
5		
Fungi	**Form**	
1		
2		
3		
4		

AIR SAMPLE

1. Sketch the macroscopic appearance of the colonies isolated from each medium.

PDA agar	Nutrient agar	Sabourad agar
◯	◯	◯

2. Describe the microscopic appearance of the different isolated microorganisms:

Bacteria	Form	Gram stain
1		
2		
3		
4		
5		
Yeast	**Form**	
1		
2		
3		
4		
5		
Fungi	**Form**	
1		
2		
3		
4		

POSTLABORATORY PROBLEMS AND QUESTIONS

1. What are the morphologic characteristics of the isolated bacterial colonies?

2. What type of microorganisms (bacteria, yeast or fungi) predominate in the different samples analyzed?

3. Why does one have to dilute the samples?

4. Why is the incubation time varied among the different types of microorganisms cultured?

Student Comments and Suggestions

Literature References

Beishir, L. *Microbiology in Practice: A Self Instructional Laboratory Course*; HarperCollins College Publishers: New York, NY, 1996.

Brown, A. E. *Benson's Microbiological Applications: Laboratory Manual in General Microbiology*; McGraw-Hill Science/Engineering/Math: New York, 2006.

Cappuccino, J.; Sherman, N. *Microbiology: A Laboratory Manual*; Benjamin Cummings: Menlo Park, CA, 2004.

Pepper, I. L; Gerba, C. P.; Brendecke, J. W. *Environmental Microbiology: A Laboratory Manual*; Academic Press: San Diego, CA, 1995.

Sharma, P. D. *Environmental Microbiology*; Alpha Science International: Harrow, U.K., 2005.

Experiment 23
Toxicity Assay Using Bacterial Growth

Reference Chapters: 9, 11

Objective

After performing the experiment, the student shall be able to:

- Use the viable cell count of a mixed bacterial culture present in spent water to assess the toxicity of sodium azide, chromium (VI), and chromium (III) as examples of inorganic toxic compounds.

Introduction

In response to the expanding stresses on the environment and in the belief that there is no single criterion by which to adequately judge the potential hazard of a given substance (either to the environment or to humans), several biological assay procedures have been developed, proposed, and used to assay toxicant impacts.

In general, there are two main groups of *in vitro* toxicity-screening tests: (a) the "health effect" tests, and (b) the "ecological effect" tests. "Health effect" toxicity tests are based on the use of subcellular components (e.g., enzymes, DNA, and RNA), isolated cells (e.g., cell cultures, red blood cells), tissue sections, or isolated whole organs. These tests consist in the determination of cell viability (e.g., plating efficiency, colony formation), cell reproduction, or macromolecular biosynthesis. "Ecological effect" tests mainly measure the acute toxicity of chemicals to aquatic organisms representing various trophic levels of the food chain.

These tests use bacteria, algae, zooplankton, benthic invertebrates, and fish for the estimation of chemical toxicity in natural and man-modified ecosystems.

In the search for rapid, relatively reproducible and inexpensive tests, bacteria appear to be sensitive sensors of chemical toxicity. This is so because they have relatively short life cycles and quick responses to changes in their environment, where they may be exposed to a wide range of toxic, organic, and inorganic compounds in natural waters, soil, and sewage treatment processes. These characteristics make bacteria suitable for the rapid screening of toxicants in natural waters.

Some screening tests are:

a) Mutagenicity test employing microorganisms and viruses, developed and standardized for determining genotoxic chemicals. The most frequently used mutagenicity test is the *Salmonella typhimurium* reverse mutation assay, which appears to be a reliable indicator of the potential carcinogenicity of organic chemicals. It uses auxotrophic mutants for histidine (his) that have either a normal or an error prone DNA repair capability and, therefore, can detect either base-pair errors or frame-shifts. Exposure to mutagenic agents induces prototrophy, and colonies of such revertants develop in histidine-free media.

b) Assay based on bacterial luminescence using *Photobacterium phosphoreum* in the Microtox® test. The Microtox® bioassay assesses acute toxicity in aquatic samples. It measures the activity of luminescent bacteria that emit light under normal metabolic conditions. Any stimulation or inhibition of their metabolism affects the intensity of the light output.

c) Assays based on the measurement of viability or growth inhibition of specific bacteria (or specific bacteria groups). Sewage microorganisms and bacteria belonging to the genera *Pseudomonas, Klebsiella, Aeromonas*, or *Citrobacter* are used for the assays.

Nonetheless, there are specific problems associated with most of these testing protocols (e.g., the choice of test organisms, inocula size, growth media, and substrate concentrations). When using bacterial cultures in toxicology testing, the decision regarding the use of pure or mixed cultures of organisms also poses a problem. Pure cultures entail fewer complications and the results are easier to interpret. If natural mixtures of microbial populations are used, the problem of deciding what sources of populations to use as inocula and how to handle and store them prior to initiating the bioassay must be considered. Pure culture testing eliminates the possibility of interspecies interactions such as synergisms, commensalisms, symbiosis, and antagonisms that occur in natural environments and that may be important for biodegradation and adaptation of the biota to the substance under study.

In order to test the toxicant capability of several substances, the viable cell count of a bacterial culture may be determined using nutrient agar, which allows for the growth of a wide variety of microorganisms. The toxicant substances are included in the medium at increasing concentrations, so that the reduction in the number of colony forming units (CFUs) is determined relative to the number grown without the toxicant.

The substances chosen for the present experiment are:

- Sodium azide (NaN_3), best known as the chemical found in automobile airbags. It is also used in detonators and other explosives. It also finds use as a chemical preservative in hospitals and laboratories, and in agriculture for pest control because azide anions prevent the function of cytochrome oxidase, an enzyme associated with respiration. Cells die as a result.

- Hexavalent chromium, Cr(VI) is a human carcinogen acting upon chronic inhalation exposures. When swallowed, it can upset the gastrointestinal tract and damage the liver and kidneys. Evidence, however, suggests that hexavalent chromium does not cause cancer when ingested, most likely because it is rapidly converted into the trivalent form, Cr(III) after entering the stomach. Chromium (VI) is a danger to human health, mainly for people who work in the steel and textile industries.

- Chromium (III) occurs naturally in many fresh vegetables, fruits, meat, grains, and yeast, and is often added to vitamins as a dietary supplement. It is an essential nutrient for humans, and shortages may cause heart conditions, metabolism disruption, and diabetes. Nevertheless, the uptake of too much Cr(III) can cause health effects (e.g., skin rashes). Chrome green is the green oxide of chromium (III), Cr_2O_3, used in enamel painting and glass staining. Trivalent chromium may cause skin irritation at high doses given parenterally but is not toxic at lower doses given orally.

Experimental Procedure

Estimated time to complete the experiment: 3 h the first day, and 30 min after 24-48 h

Materials	Reagents
3 500-mL Erlenmeyer flasks	Wastewater sample
6 1-mL pipets	Saline isotonic solution
20 sterile Petri dishes,	(0.9% NaCl)
60 × 15 mm	Nutrient agar
10 1-mL pipets	
10 10-mL pipets	
10 screw cap 13 × 100 mm	
test tubes	
Cotton caps	
Incubators (37°C, 30°C)	
Autoclave	
1 mixing plate	
1 Bunsen burner	
Test tube rack	
Inoculating loop	
Glass slides and cover slips	
Sterile spatula	

When doing microbiological laboratory work, the following rules must be followed:

1. Use a protective garment (e.g., a lab coat).
2. Never eat or drink in the laboratory, and avoid placing objects in your mouth.
3. Always wash your hands before leaving the lab.
4. If you spill living organisms, cover the spilled material with paper towels and pour a laboratory disinfectant over the towels and the entire contaminated area. Wait 15 min before you clean it up.

Preparation of the agar medium

Nutrient agar is prepared in the concentration indicated by the manufacturer. Mix thoroughly. Gently heat and bring the mixture to boil. Autoclave for 15 min at 15 psi and 121°C. Maintain the agar at 45°C until it is poured in sterile Petri dishes.

Autoclave ten 10-mL screw cap tubes for each toxicant tested, 100 mL of saline solution (0.9% NaCl) and 1- and 10-mL pipets.

Preparation of the control plates

A small sample of spent water can be used (e.g., the effluent of a wastewater treatment plant). Dilute 1 mL of this sample to 1/10, 1/100, and 1/1000 by using the screw cap tubes in the following way:

• Add 9 mL of sterile saline solution to each tube.
• Add 1 mL of the sample to the first tube. Mix thoroughly. This is a 10^{-1} dilution. Transfer 1 mL of this solution to the second tube and repeat the procedure to prepare two additional dilutions (10^{-2} and 10^{-3}).

Use an aseptic technique when making serial dilutions and plating.

Always use a clean, sterile pipet for all transfers.

Preparation of the samples

Prepare small Petri dishes in the following way for each of the proposed toxicants.

Petri dish no. Toxicant (conc.), Vol.		1	2	3	4	5	6	7	8	9
Sodium azide (2 mg/mL)	(µL)	0	25	50	75	100	25	50	75	100
Aseptic saline solution	(µL)	100	75	50	25	0	75	50	25	0
Diluted sample, 10^{-2}	(µL)	100	100	100	100	100	-	-	-	-
Diluted sample, 10^{-3}	(µL)	-	-	-	-	-	100	100	100	100
Nutrient agar medium	(mL)	10	10	10	10	10	10	10	10	10

Petri dish no. Toxicant (conc.), Vol.	1	2	3	4	5	6	7	8	9	10	11
$K_2Cr_2O_7$ (29.4 mg/ 10 mL = 0.02 mmol Cr^{6+}/mL) (µL)	0	25	50	75	100	200	25	50	75	100	200
Aseptic saline solution (µL)	200	175	150	125	100	0	175	150	125	100	0
Diluted sample, 10^{-2} (µL)	100	100	100	100	100	100	-	-	-	-	-
Diluted sample, 10^{-3} (µL)	-	-	-	-	-	-	100	100	100	100	100
Nutrient agar medium (mL)	10	10	10	10	10	10	10	10	10	10	10

Petri dish no. Toxicant (conc.), Vol.	1	2	3	4	5	6	7	8	9	10	11
Cr_2O_3 (15.2 mg/10 mL = 0.02 mmol Cr^{3+}/mL) (µL)	0	25	50	75	100	200	25	50	75	100	200
Aseptic saline solution (µL)	200	175	150	125	100	0	175	150	125	100	0
Diluted sample, 10^{-2} (µL)	100	100	100	100	100	100	-	-	-	-	-
Diluted sample, 10^{-3} (µL)	-	-	-	-	-	-	100	100	100	100	100
Nutrient agar medium (mL)	10	10	10	10	10	10	10	10	10	10	10

Prepare each Petri dish in duplicate.

- Reagents are added first and the agar is added at 45°C, before it solidifies. Move with slow horizontal movements to homogenize. Let the mixture stand still for 20 minutes.
- Incubate at 30°C for 24–48 h.
- Count the number of colonies in each plate. Determine the number of colony forming units (CFU)/mL in the control plates and in each dilution (take the average of two or three determinations).

Autoclave all Petri dishes when the experiment is concluded to kill the isolated microorganisms

Name_____Section_____Date_____

Instructor_____Partner_____

PRELABORATORY REPORT SHEET—EXPERIMENT 23

Objectives

Flow sheet of procedure

Waste containment and recycling procedure

PRELABORATORY QUESTIONS AND PROBLEMS

1. What is *a colony forming unit* and why is this different from the number of bacteria present in a sample?

2. The information on chromium (VI) given in the Introduction refers to the toxicity of this element to humans. Investigate why it is toxic to microorganisms. Can this explanation be extrapolated to humans?

3. To compare the toxicities of the different substances in this experiment, what variables should be kept constant?

Additional Related Projects

- Instead of a sample of a wastewater treatment plant effluent, any spent water may be used. It is only necessary to establish the dilution range in which the aerobic microorganisms may be plate-counted.

- Test other toxicants. It is only necessary to establish the concentration in which the growth inhibition of the mixed culture can be evaluated.

Name_____Section_____Date_____

Instructor_____Partner_____

LABORATORY REPORT SHEET—EXPERIMENT 23

Number of CFU/mL in the control plate (plate **1**) = N_0
Number of CFU/mL in each dilution plate = N_t
Plot ln N_0/N_t vs toxicant concentration and compare the toxicity of each substance.

Toxicant concentration

POSTLABORATORY PROBLEMS AND QUESTIONS

1. Was the concentration of each substance adequate to observe its toxic effect?

2. Order the substances employed in this experiment by their increasing toxic effect.

3. What interpretation can you give to the slope of each curve obtained in your graph?

Student Comments and Suggestions

Literature References

Bitton, G.; Dotks, J. *Toxicity Testing Using Microorganisms;* CRC Press: Boca Raton, FL, 1986.

Experiment 24
Wastewater Disinfection

Reference Chapters: 10, 11

Objectives

After performing the experiment, the student shall be able to:

- Evaluate the efficiency of different physical and chemical disinfecting agents (sodium hypochlorite, ozone, ultraviolet light, and titanium dioxide).
- Become familiar with the use of *coliform counting plates*.
- Perform a kinetic analysis on the rate of coliform inactivation that each agent causes.

Introduction

Disinfection is the *selective* destruction of pathogenic organisms, whereas *sterilization* implies the *complete* destruction of all microorganisms. Disinfection is used during wastewater treatment in order to reduce pathogens to an acceptable level, which helps in the control of diseases caused by contaminated water and food-stuffs by bacteria and other microorganisms.

Methods used for wastewater treatment may be physical (e.g., heat), chemical (e.g., chlorine and its derivates, ozone, hydrogen peroxide, and colloidal silver), mechanical (e.g., sedimentation and filtration), or radiative (e.g., electromagnetic and acoustic). The efficiency of a disinfecting agent depends on several variables such as the contact time with pathogens in the sample, their concentration, temperature, number, and the nature of the liquid in which the pathogens are suspended. One of the main

variables is the contact time. In general, the longer the contact time with a fixed concentration of disinfectant, the higher the mortality. First-order kinetics is usually followed. This relationship is known as the *Chick equation*:

$$\frac{dN}{dt} = -kN_t \qquad (1)$$

where N_t = number of microorganisms at time t, and k = constant (in s^{-1}). If N_0 is the number of microorganism at $t = 0$, the integrated equation is:

$$\ln \frac{N_t}{N_0} = -kt \qquad (2)$$

It is common to find deviations from first-order kinetics caused by the resistance of subpopulations (within a mixed population) to the disinfecting agent or to the presence of protecting factors in the medium, both of which interfere with pathogen destruction. This effect can be evaluated by assuming that the mortality rate under different conditions follows the relationship:

$$\ln \frac{N_t}{N_0} = -kt^m \qquad (3)$$

where m is a constant. If $m < 1$, the mortality rate *decreases* with time, while if $m > 1$, it *increases* with time. The value of m can be determined by plotting the equation:

$$\log\left(-\ln \frac{N_t}{N_0}\right) = \log k + m \log t \qquad (4)$$

A brief overview of the disinfectants that will be used in this experiment is now given. More extensive discussions are given in Chapter 10.

Hypochlorite

When chlorine gas is dissolved in water it reacts as follows:

$$Cl_{2(g)} + H_2O_{(l)} \rightarrow HOCl + H^+ + Cl^-$$
$$HOCl \rightarrow H^+ + OCl^-$$

When sodium hypochlorite is used, the reaction is:

$$NaOCl + H_2O_{(l)} \rightarrow HOCl + NaOH$$

The $HOCl/OCl^-$ ratio depends on pH. As discussed in Chapter 10, the amount of chlorine present in the sample is the *available free chlorine*. Hypochlorous acid is a better disinfectant agent than its anion. Pathogenic bacteria are easily attacked, although resistance increases in the order: vegetative bacteria < protozoan cysts < helminth eggs.

The presence of compounds that react with chlorine boosts the required quantity for adequate disinfection. In addition, toxic secondary products can be formed, and therefore new disinfectants have been developed.

Ozone

Ozone is an unstable gas that decomposes rapidly to form oxygen. As a result, it is synthesized *in situ*. Its production methods are reviewed in Chapter 10. This disinfecting agent inactivates pathogens, oxidizes Fe(II) and manganese (II) ions, combats wastewater odors and color, and oxidizes refractory organics and trihalomethanes. Its efficiency is not influenced by pH and does not interfere with ammonia. In general it does not leave residual compounds in water.

Ozone is a better oxidizing agent than chlorine; its minimum concentration is 0.1 mg/L for bacteria inactivation and 0.5 g/L for protozoan cysts under appropriate exposure times. Its disinfecting action is due to the production of free radicals, which affect cellular permeability, enzymatic activity and the DNA.

Ultraviolet Light

Ultraviolet radiation is an efficient bactericide and virucide, as it reacts with DNA at 260 nm, although the time for treatment depends on the type of microorganism present. Low-pressure mercury arc lamps are typically used, as they generate 85% monochromatic radiation at 253.7 nm, which is within the optimum range for germicide action.

Suspended particles and colored substances may interfere with the process; that is why UV disinfection is especially useful for the treatment of potable water. It has the advantage that it does not form toxic secondary products. Challenges include a practical difficulty for the determination of the optimum dose; the lamps require constant maintenance, and its cost is higher than that of a chlorine treatment.

Titanium dioxide +UV

Catalytic disinfection is a modern alternative that involves the use of water-soluble or water-suspendable sensitizers to generate transient oxygen species.

Solar radiation promotes coliform and virus photocatalytic disinfection. This involves oxidizing reactions in which electron removal and the generation of hydroxyl radicals attack biological molecules at rates controlled by their diffusion through the walls and membranes of microorganisms.

In the present experiment, the disinfection treatments discussed above are applied to wastewater samples containing mixed populations of microorganisms, and the surviving colonies are counted.

Experimental Procedure

Estimated time to complete the experiment: 3 h the first day and 30 min after 24–48 h.

Materials	Reagents
4 10-mL beakers	wastewater sample
3 20-mL beakers	4.5 ppm NaOCl
10 screw cap test tubes	TiO_2
20 sterile Petri dishes, 60 × 15 mm (if McConkey agar is used)	coliform counting plates or McConkey agar medium
Chronometer	Saline isotonic solution, (NaCl 0.9%)
10 sterilized 1-mL pipets	
Commercial ozonizer	
UV lamp	
Incubator	
Test tube rack	

When doing microbiological laboratory work, the following rules must be followed:

1. Use a protective garment (e.g., a lab coat).
2. Never eat or drink in the laboratory, and avoid placing objects in your mouth.
3. Always wash your hands before leaving the lab.
4. If you spill living organisms, cover the spilled material with paper towels and pour a laboratory disinfectant over the towels and the entire contaminated area. Wait 15 min before you clean it up.

After fixed time intervals of exposure to a given disinfection agent, 1-mL wastewater aliquots are spread on coliform plaques (e.g., 3M Petrifilm™). The use of these plaques avoids the preparation of a sterilized culture medium specific for coliform microorganisms. (Instead of the plaques, McConkey agar medium can be used and sterilized according to the manufacturer instructions, and Petri dishes prepared). McConkey agar contains crystal violet, which is inhibitory to Gram-positive bacteria, and allows the isolation of Gram-negative bacteria. Incorporation of the carbohydrate lactose, bile salts, and the pH indicator neutral red permits differentiation of enteric bacteria on the basis of their ability to ferment lactose. Coliform bacilli produce acid as a result of lactose fermentation and exhibit a red coloration).

The plaques are incubated for 24 hours at 37°C, and the resulting red-colored colonies are counted. There should be fewer than 50 colonies per plaque; if not, dilute the samples before spreading them onto the plaques. If dilution is necessary prepare a dilution series as follows:

1. Take 1 mL of the sample and mix with 9 mL of saline solution in a sterile screw cap test tube; mix thoroughly to have a 10^{-1} dilution.
2. Transfer 1 mL of the 10^{-1} dilution to another test tube containing 9 mL of the saline solution, using a sterile pipet and repeat the step in order to have 10^{-2} and 10^{-3} dilutions. Mix thoroughly.

Use an aseptic technique when making serial dilutions and plating. Always use a clean, sterile pipet for all transfers.

Hypochlorite Treatment

Take five 5-mL wastewater samples in 10-mL beakers. Add 1 mL of a 4.5 ppm NaOCl solution to beakers 2, 3, 4, and 5. Make sure that the final concentration of NaOCl is less than 2 ppm. Stir at constant speed. Take a 1-mL aliquot from beaker 1 at time 0 (this is the control), a similar aliquot from beaker 2 after 5 min, another aliquot from beaker 3 after 10 min, and so on up to 20 min. Place each aliquot on a plaque, incubate, and count the red colonies as described above. If Petri dishes with McConkey agar were used, a 0.1 mL aliquot of each dilution can be used.

Ozone Treatment

Take a 10-mL wastewater sample in a 20-mL beaker. A commercial ozonizer with a porous filter can be used to deliver small ozone bubbles to each sample. (For instance, one can use a BIOZON FAGOR ozonizer that produces 4 mg of O_3/min). Take a 1-mL aliquot from the beaker after 0, 30, 90 and 180 s, and place each aliquot on a plaque. Incubate and count as described above.

UV Treatment

Take a 10-mL wastewater sample in a 20-mL beaker and place it in a dark compartment. Irradiate the sample with a 254-nm, 9-W lamp placed at 12 cm above it. Stir at constant speed. Take a 1-mL aliquot from the beaker after 0, 60, 150, and 240 s, and place each aliquot on a plaque. Incubate and count as described above.

TiO_2 + UV Treatment

Take a 10-mL wastewater sample in a 20-mL beaker and add 20 mg of TiO_2 (the anatase phase) to form a suspension. Stir at a constant speed. Place the experimental set-up in a dark compartment. Irradiate the sample with a 254-nm, 9-W lamp placed at 12 cm above it. Take a 1-mL aliquot from the beaker after 0, 120, 300, and 480 s, and place each aliquot on a plaque. Incubate and count as described above.

Autoclave all Petri dishes when the experiment is concluded to kill the isolated microorganisms.

Name_____Section_____Date_____

Instructor_____Partner_____

PRELABORATORY REPORT SHEET—EXPERIMENT 24

Objectives

Flow sheet of procedure

Waste containment and recycling procedure

PRELABORATORY QUESTIONS AND PROBLEMS

1. What is the difference between *disinfection* and *sterilization*?
2. What does the constant k in the integrated form of the Chick equation represent?
3. What does *available free chlorine* mean?
4. The chlorine disinfection treatment uses a 4.5 ppm NaOCl solution. It can be prepared from a 4–5% commercial solution. Assuming that this last concentration is 4.5%, indicate the amount of the commercial solution that must be diluted in order to prepare 5 mL of a 4.5 ppm solution.
5. How does the disinfection mechanism differ when using ultraviolet radiation alone vs using this radiation with titanium oxide?

ADDITIONAL RELATED PROJECT

• Use the same methodology as in this experiment to evaluate the dependence of microorganism disinfection rate with the disinfectant or photocatalyst concentration, light intensity, temperature, or other variables that may affect this process.

Name_____Section_____Date_____

Instructor_____Partner_____

LABORATORY REPORT SHEET—EXPERIMENT 24

a) Count the number of colony-forming units (CFUs) on each plaque and fill the following table.

	Time, s	N_0	N_t	$\ln N_t/N_0$	$\log (-\ln N_t/N_0)$
Hypochlorite					
Ozone					
UV					
TiO$_2$ + UV					

b) Plot $\ln N_t/N_0$ vs t.

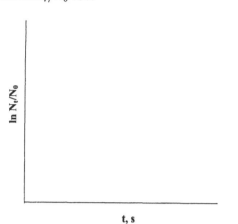

c) Plot $\log(-\ln N_t/N_0)$ vs $\log t$.

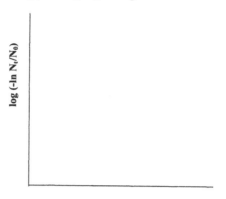

POSTLABORATORY PROBLEMS AND QUESTIONS

1. Does any disinfectant in this experiment display first-order kinetics?
2. Is the value of m for any disinfectant in this experiment significantly different than the others? If so, what can this mean?
3. According to the results of your experiment, which is the best disinfecting agent for the sample used?

Student Comments and Suggestions

Literature References

Bitton, G. *Wastewater Microbiology*; Wiley-Liss: New York, 1994.

Metcalf & Eddy. *Wastewater Engineering. Treatment, Disposal, Reuse*; Mc Graw-Hill International Editions: New York, 1991.

Rajeshwar, K.; Ibanez, J. G. "Electrochemical Aspects of Photocatalysis: Application to Detoxification and Disinfection Scenarios," *J. Chem. Ed.* **1995**, *72*, 1044–1049.

Tyrrell, Sh. A.; Rippey, S. R.; Watkins, W. D. "Inactivation of Bacterial and Viral Indicators in Secondary Sewage Effluents, Using Chlorine and Ozone," *Wat. Res.* **1995**, *29*, 2483–2490.

Watts, R. J.; Kong, S.; Orr, M. P.; Miller, G. C.; Henry, B. E. "Photocatalytic Inactivation of Coliform Bacteria and Viruses in Secondary Wastewater Effluents," *Wat. Res.* **1995**, *29*, 95–100.

Appendix
Analytical Environmental Chemistry Experiments in the Literature

Adami, G. "A New Project-Based Lab for Undergraduate Environmental and Analytical Chemistry," *J. Chem. Educ.* **2006**, *83*, 253–256.

Anderson, C. P.; Saner, W. B. "A Practical Experiment for Determining a Pervasive, Persistent Pollutant," *J. Chem. Educ.* **1984**, *61*, 738–739.

Allen, H. C.; Brauers, T.; Finlayson-Pitts, B. J. "Illustration of Deviations in the Beer-Lambert Law in an Instrumental Analysis Laboratory: Measuring Atmospheric Pollutants by Differential Optical Absorption Spectrometry," *J. Chem. Educ.* **1997**, *74*, 1459–1462.

Amey, R. L. "Atmospheric Smog Analysis in a Balloon Using FTIR Spectroscopy—A Novel Experiment for the Introductory Laboratory," *J. Chem. Educ.* **1992**, *69*, A148–151.

Arnold, R. J. "The Water Project: A Multi-Week Laboratory Project for Undergraduate Analytical Chemistry," *J. Chem. Educ.* **2003**, *80*, 58–60.

Atterholt, C.; Butcher, D. J.; Bacon, J. R.; Kwochka, W. R.; Woosley, R. "Implementation of an Environmental Focus in an Undergraduate Chemistry Curriculum by the Addition of Gas Chromatography-Mass Spectrometry," *J. Chem. Educ.* **2000**, *77*, 1550–1573.

Baird, M. J. "Analysis of an Air Conditioning Coolant Solution for Metal Contamination Using Atomic Absorption Spectroscopy. An Undergraduate Instrumental Analysis Exercise Simulating an Industrial Assignment," *J. Chem. Educ.* **2004**, *81*, 259–261.

Balko, B. A.; Tratnyek, P. G. "A Discovery-Based Experiment Illustrating How Iron Metal Is Used to Remediate Contaminated Groundwater," *J. Chem. Educ.* **2001**, *78*, 1661–1664.

Bell, S. The Role of Automated Instrumentation in Undergraduate Chemistry," *J. Chem. Educ.* **2000**, *77*, 1624–1626.

Bendikov, T. A.; Harmon, T. C. "A Sensitive Nitrate Ion-Selective Electrode from a Pencil Lead. An Analytical Laboratory Experiment," *J. Chem. Educ.* **2005**, *82*, 439–441.

Bower, N. W. "Environmental Chemical Analysis (by B. B. Kebbekus and S. Mitra)," *J. Chem. Educ.* **1999**, *76*, 1489–1490.

Breslin, V. T.; Sanudo-Wilhelmy, S. A. "The Lead Project. An Environmental Instrumental Analysis Case Study," *J. Chem. Educ.* **2001**, *78*, 1647–1651.

Bruce, D.; Kuhn, A.; Sojic, N. "Electrochemical Removal of Metal Cations from Wastewater Monitored by Differential Pulse Polarography," *J. Chem. Educ.* **2004**, *81*, 255–258.

Buell, P.; Girard, J. *Laboratory Manual for Chemistry Fundamentals*, 2nd *ed.*; Jones and Bartlett: Boston, MA, 2002.

- Concentration Units for Environmental Sampling (Experiment 9)
- Determination of Lead in Paint Chips (Experiment 20)
- Environmental Sampling (Experiment 22)

Buffin, B. P. "Removal of Heavy Metals from Water: An Environmentally Significant Atomic Absorption Spectrometry Experiment," *J. Chem. Educ.* **1999**, *76*, 1678–1679.

Burke, J. R. (Waterose Environmental). "The Colorimetric Analysis of Nitrates and Phosphates in Surface Ponds and Streams," http://www.geocities.com/rainforest/vines/4301/lab05.html

Butala, S. J.; Zarrabi, K.; Emerson, D. W. "Sampling and Analysis of Lead in Water and Soil Samples on a University Campus—A Student Research Project," *J. Chem. Educ.* **1995**, *72*, 441–444.

Cancilla, D. A. "Integration of Environmental Analytical Chemistry with Environmental Law: The Development of a Problem-Based Laboratory," *J. Chem. Educ.* **2001**, *78*, 1652–1160.

Cavinato, A. G.; Kelley, R. B. "Integrating Service Learning in the Environmental Chemistry Laboratory," 225th. Am. Chem. Soc. National Meeting, New Orleans, LA, March 23–27, 2003. Chem. Ed. Division Paper # 87.

Chang, J. C.; Levine, S. P.; Simmons, M. S. "A Laboratory Exercise for Compatibility Testing of Hazardous Wastes in an Environmental Analysis Course," *J. Chem. Educ.* **1986**, *63*, 640–643.

Chasteen, T. G. *Qualitative and Instrumental Analysis of Environmentally Significant Elements*, 1st ed.; Wiley: New York, 1993.

Collado-Sánchez, C.; Hernández-Brito, J. J.; Pérez-Peña, J.; Torres-Padrón, M. E; Gelado-Caballero, M. D. "Adsorptive Stripping Voltammetry of Environmental Indicators: Determination of Zinc in Algae," *J. Chem. Educ.* **2005**, *82*, 271–273.

Correia, P. R. M.; Oliveira, P. V. "Simultaneous Atomic Absorption Spectrometry for Cadmium and Lead Determination in Wastewater. A Laboratory Exercise," *J. Chem. Educ.* **2004**, *81*, 1174–1176.

Correia, P. R. M.; Siloto, R. C.; Cavicchioli, A.; Oliveira, P. V.; Rocha, F. R. P. "Green Analytical Chemistry in Undergraduate Laboratories: Flow-Injection Determination of Creatinine in Urine with Photochemical Treatment of Waste," *Chem. Educ.* **2004**, *9* (4), 242–246.

Crisp, G. T.; Williamson, N. M. "Separation of Polyaromatic Hydrocarbons Using 2-Dimensional Thin-Layer Chromatography: An Environmental Chemistry Experiment," *J. Chem. Educ.* **1999**, *76*, 1691–1692.

Csuros, M. *Environmental Sampling and Analysis: Lab Manual*; CRC Lewis Publishers: Boca Raton, FL, 1997

Daniel, D. W. V.; Allen, A. G.; Brown, G. R.; Chandrupatta, H.; Fowler, S. A.; Richmond, D. J.; Skipper, S. W.; Taylor, J. A.; Wasiak, A. E. "A Simple UV Experiment of Environmental Significance," *J. Chem. Educ.* **1994**, *71*, 83.

De Bruyn, W. J. "Incorporating Hands-on Environmental Chemistry Research Projects into General Education Science Courses," 225th. Am. Chem. Soc. National Meeting, New Orleans, LA, March 23–27, 2003. Chem. Ed. Division Paper # 84.

Demay, S.; Martin-Girardeau, A.; Gonnord, M.-F. "Capillary Electrophoretic Quantitative Analysis of Anions in Drinking Water," *J. Chem. Educ.* **1999**, *76*, 812–815.

Dinardi, S. R.; Briggs, E. S. "Hydrocarbons in Ambient Air—A Laboratory Experiment," *J. Chem. Educ.* **1975**, *52*, 811–822.

Donahue, C. J.; D'Amico, T.; Exline, J. A. "Synthesis and Characterization of a Gasoline Oxygenate, Ethyl *tert*-Butyl Ether," *J. Chem. Educ.* **2002**, *79*, 724–726.

Draper, A. J. "Integrating Project-Based Service-Learning into an Advanced Environmental Chemistry Course," *J. Chem. Educ.* **2004**, *81*, 221–224.

Dunnivant, F. M. *Environmental Laboratory Exercises for Instrumental Analysis and Environmental Chemistry*; Wiley-Interscience: New York, 2004.

1. How to Keep a Legally Defensible Laboratory Notebook.
2. Statistical Analysis.
3. Field Sampling Equipment for Environmental Samples.
4. Determination of Henry's Law Constants.
5. Global Warming: Determining If a Gas Is Infrared Active.
6. Monitoring the Presence of Hydrocarbons in Air around Gasoline Stations.
7. Determination of an Ion Balance for a Water Sample.
8. Measuring the Concentration of Chlorinated Pesticides in Water Samples.
9. Determination of Chloride, Bromide, and Fluoride in Water Samples.
10. Analysis of Nickel Solutions by Ultraviolet–Visible Spectrometry.
11. Determination of the Composition of Unleaded Gasoline Using Gas Chromatography.
12. Precipitation of Metals from Hazardous Waste.
13. Determination of the Nitroaromatics in Synthetic Wastewater from a Munitions Plant.
14. Determination of a Surrogate Toxic Metal in a Simulated Hazardous Waste Sample.

15. Reduction of Substituted Nitrobenzenes by Anaerobic Humic Acid Solutions.

16. Soxhlet Extraction and Analysis of a Soil or Sediment Sample Contaminated with *n*-Pentadecane.

17. Determination of a Clay–Water Distribution Coefficient for Copper.

18. Determination of Dissolved Oxygen in Water Using the Winkler Method.

19. Determination of the Biochemical Oxygen Demand of Sewage Influent.

20. Determination of Inorganic and Organic Solids in Water Samples: Mass Balance Exercise.

21. Determination of Alkalinity of Natural Waters.

22. Determination of Hardness in a Water Sample.

23. pC–pH Diagrams: Equilibrium Diagrams for Weak Acid and Base Systems.

24. Fate and Transport of Pollutants in Rivers and Streams.

25. Fate and Transport of Pollutants in Lake Systems.

26. Fate and Transport of Pollutants in Groundwater Systems.

27. Transport of Pollutants in the Atmosphere.

28. Biochemical Oxygen Demand and the Dissolved Oxygen Sag Curve in a Stream: Streeter–Phelps Equation.

Dunnivant, F. M. "Analytical Problems Associated with the Analysis of Metals in a Simulated Hazardous Waste," *J. Chem. Educ.* **2002**, *79*, 718–720.

Elrod, M. J. "Greenhouse Warming Potentials from the Infrared Spectroscopy of Atmospheric Gases," *J. Chem. Educ.* **1999**, *76*, 1702–1705.

Evans, J. J. "Turbidimetric Analysis of Water and Wastewater Samples Using a Spectrofluorimeter," *J. Chem. Educ.* **2000**, *77*, 1609–1611.

Fleurat-Lessard, P.; Pointet, K.; Renou-Gonnord, M. F. "Quantitative Determination of PAHs in Diesel Engine Exhausts by GC-MS", *J. Chem. Educ.* **1999**, *76*, 962–965.

Garizi, N.; Macias, A.; Furch, T.; Fan, R.; Wagenknecht, P.; Singmaster, K. A. "Cigarette Smoke Analysis Using an Inexpensive Gas-Phase IR Cell," *J. Chem. Educ.* **2001**, *78*, 1665–1666.

Giokas, D. L.; Paleologos, E. K.; Karayannis, M. I. "Micelle-Mediated Extraction of Heavy Metals from Environmental Samples: An Environmental Green Chemistry Laboratory Experiment," *J. Chem. Educ.* **2003**, *80*, 61–64.

Glover, I. T.; Johnson, F. T. "Determination of Nitrite in Meat Samples Using a Colorimetric Method," *J. Chem. Educ.* **1973**, *50*, 426–427.

Glover, I. T.; Minter, A. P. "Analysis of Chlorinated Hydrocarbon Pesticides—An Experiment for Non-Science Majors," *J. Chem. Educ.* **1974**, *51*, 685–686.

Goebel, A.; Vos, T. Louwagie, A.; Lundbohm, L.; Brown, J. H. "Lead-Testing Service to Elementary and Secondary Schools Using Anodic Stripping Voltammetry," *J. Chem. Educ.* **2004**, *81*, 214–217.

Gruenhagen, J. A.; Delaware, D.; Ma, Y. "Quantitative Analysis of Non-UV-Absorbing Cations in Soil Samples by High-Performance Capillary Electrophoresis. An Experiment for Undergraduate Instrumental Analysis Laboratory," *J. Chem. Educ.* **2000**, *77*, 1613–1616.

Guisto-Norkus, R.; Gounili, G.; Wisniecki, P.; Hubball, J. A; Smith, S. R.; Stuart, J. D. "An Environmentally Significant Experiment Using GC/MS and GC Retention Indices in an Undergraduate Analytical Laboratory," *J. Chem. Educ.* **1996**, *73*, 1176–1178.

Hage, D. S.; Chattopadhyay, A.; Wolfe, C. A. C.; Grundman, J.; Kelter, P. B. "Determination of Nitrate and Nitrite in Water by Capillary Electrophoresis: An Undergraduate Laboratory Experiment," *J. Chem. Educ.* **1998**, *75*, 1588–1590.

Hall, S.; Reichardt, P. B. "DDE Levels in Birds: An Environmentally Oriented Undergraduate Experiment," *J. Chem. Educ.* **1974**, *51*, 684.

Hauser, B. A. *Practical Manual of Wastewater Chemistry;* Ann Arbor Press: Chelsea, MI, 1996.

Herrera-Melian, J. A.; Dona-Rodriguez, J. M.; Hernandez-Brito, J.; Pena, J. P. "Voltammetric Determination of Ni and Co in Water Samples," *J. Chem. Educ.* **1997**, *74*, 1444–1445.

Hope, W. W.; Johnson, C.; Johnson, L. P. "Tetraglyme Trap for the Determination of Volatile Organic Compounds in Urban Air. Projects for Undergraduate Analytical Chemistry," *J. Chem. Educ.* **2004**, *81*, 1182–1186.

Hope, W. W.; Johnson, L. P. "Urban Air: Real Samples for Undergraduate Analytical Chemistry," *Anal. Chem.* **2000**, *72* (13), 460A–467A.

Jaffe, D.; Herndon, S. "Measuring Carbon Monoxide in Auto Exhaust by Gas-Chromatography," *J. Chem. Educ.* **1995**, *72*, 364–366.

Jarosch, R. "The Determination of Pesticide Residues—A Laboratory Experiment," *J. Chem. Educ.* **1973**, *50*, 507–508.

Jenkins, J. D.; Orvis, J. N.; Smith, C. J.; Manley, C.; Rice, J. K. "Including Non-Traditional Instrumentation in Undergraduate Environmental Chemistry Courses," *J. Chem. Educ.* **2004**, *81*, 22–23.

John, R.; Lord, D. "Determination of Anionic Surfactants Using Atomic Absorption Spectrometry and Anodic Stripping Voltammetry," *J. Chem. Educ.* **1999**, *76*, 1256–1258.

Kegley, S. E.; Hansen, K. J.; Cunningham, K. L. "Determination of Polychlorinated Biphenyls (PCBs) in River and Bay Sediments: An Undergraduate Laboratory Experiment in Environmental Chemistry Using Capillary Gas Chromatography with Electron Capture Detection," *J. Chem. Educ.* **1996**, *73*, 558–562.

Kegley, S.; Stacy, A. M.; Carroll, M. K. "Environmental Chemistry in the General Chemistry Laboratory, Part I: A Context-Based Approach To Teaching Chemistry,"*Chem. Educ.* **1996**, *1*, 1–14.

Kegley, S. E.; Wise, L. J. *Pesticides in Fruits and Vegetables;* University Science Books: Sausalito, CA, 1998.

Kesner, L.; Eyring, E. M. "Service-Learning General Chemistry: Lead Paint Analyses," *J. Chem. Educ.* **1999**, *76*, 920–923.

Krasnoperov, L. N.; Stepanov, V. "Introduction of Laser Photolysis–Transient Spectroscopy in an Undergraduate Physical Chemistry Laboratory: Kinetics of Ozone Formation," *J. Chem. Educ.* **1999**, *76*, 1182–1183.

Lambert, J. L.; Meloan, C. E. "A Simple Qualitative Analysis Scheme for Several Environmentally Important Elements," *J. Chem. Educ.* **1977**, *54*, 249–252.

Libes, S. "Constructing Environmental Impact Statements. An Organizational Focus for Teaching Analytical Environmental Chemistry," *J. Chem. Educ.* **1999**, *76*, 1649–1656.

Libes, S. M. "Learning Quality Assurance/Quality Control Using U.S. EPA Techniques: An Undergraduate Course for Environmental Chemistry Majors," *J. Chem. Educ.* **1999**, *76*, 1642–1648.

Lieu, V. T.; Cannon, A.; Huddeston, W. E. "A Non-Flame Atomic Absorption Attachment for Trace Mercury Determination," *J. Chem. Educ.* **1974**, *51*, 752–753.

Lodge, K. B. "Use of an Emission Analyzer to Demonstrate Basic Principles," *Chem. Eng. Educ.* **2000**, *34* (2), 178–184.

Lyons, R. G.; Crossley, P. C.; Fortune, D. "High-Sensitivity Gamma Radiation Monitor for Teaching and Environmental Applications," *J. Chem. Educ.* **1994**, *71*, 524–527.

Mabury, S. A.; Mathers, D.; Ellis, D. A.; Lee, P.; Marsella, A. M.; Douglas, M. "An Undergraduate Experiment for the Measurement of Trace Metals in Core Sediments by ICP-AES and GFAAS," *J. Chem. Educ.* **2000**, *77*, 1611–1612.

Marsella, A. M.; Huang, J.; Ellis, D. A.; Mabury, S. A. "An Undergraduate Field Experiment for Measuring Exposure to Environmental Tobacco Smoke in Indoor Environments," *J. Chem. Educ.* **1999**, *76*, 1700–1701.

Marshall, D. R.; Owen, N. L.; Underhill, A. E. "Analysis of Industrial Waste. A Quantitative Laboratory Project", *J. Chem. Educ.* **1977**, *54*, 584.

McGowin, A. E.; Hess, G. G. "Incorporation of GC-MS into an Environmental Science Curriculum," *J. Chem. Educ.* **1999**, *76*, 23–64.

Medhurst, L. J. "FTIR Determination of Pollutants in Automobile Exhaust: An Environmental Chemistry Experiment Comparing Cold-Start and Warm-Engine Conditions," *J. Chem. Educ.* **2005**, *82*, 278–281.

Mehra, M. C. "Radiometric Analysis of Ammonia in Water," *J. Chem. Educ.* **1972**, *49*, 837–838.

Miguel, A. H.; Braun, R. D. "Fluorimetric Analysis of Nitrate in Real Samples," *J. Chem. Educ.* **1974**, *51*, 682–683.

Mihok, M.; Keiser, J. T.; Bortiatynski, J. M.; Mallouk, T. E. "An Environmentally-Focused General Chemistry Laboratory," *J. Chem. Educ.* **2006**, *83*, 250–252.

• Iron and Alkalinity Determinations (Experiment 2)
• Atomic Absorption Cation Determination (Mg and Ca) (Experiment 3)
• Paper and Liquid Chromatography (Experiment 4)
• Ion Chromatography (Experiment 5)

Myrick, M. L.; Greer, A. E.; Nieuwland, A.; Priore, R. J.; Scaffidi, J.; Andreatta, D.; Colavita, P. "Fine-Structure Measurements of Oxygen A Band Absorbance for Estimating the Thermodynamic Average Temperature of the Earth's Atmosphere. An Experiment in Physical and

Environmental Chemistry," *J. Chem. Educ.* **2006**, *83*, 263–264.

Nahir, T. M.."Analysis of Semivolatile Organic Compounds in Fuels Using Gas Chromatography-Mass Spectrometry," *J. Chem. Educ.* **1999**, *76*, 1695–1696.

Norkus, R. G.; Gounii, G.; Wisniecki, P.; Hubball, J. A.; Smith, S, R.; Stuart, J. D. "An Environmentally Significant Experiment Using GC/MS and GC Retention Indices in an Undergraduate Analytical Laboratory," *J. Chem. Educ.* **1996**, *73*, 1176–1178

Notestein, J.; Helias, N.; Wentworth, W. E.; Dojahn, J. G.; Chen, E. C. M.; Stearns, S. D. "A Unique Qualitative GC Experiment for an Undergraduate Instrumental Methods Course Using Selective Photoionization Detectors," *J. Chem. Educ.* **1998**, *75*, 360–364.

O'Hara, P. B.; Sanborn, J. A.; Howard, M. "Pesticides in Drinking Water: Project-Based Learning within the Introductory Chemistry Curriculum," *J. Chem. Educ.* **1999**, *76*, 1673–1679.

Olson, T. M.; Gonzalez, A. C.; Vasquez, V. R. "Gas Chromatography Analyses for Trihalomethanes: An Experiment Illustrating Important Sources of Disinfection By-Products in Water Treatment," *J. Chem. Educ.* **2001**, *78*, 1231–1234.

Ondrus, M. G. "The Determination of NO_x and Particulate in Cigarette Smoke—A Student Laboratory Experiment," *J. Chem. Educ.* **1979**, *56*, 551–552.

Persinger, J. D.; Hoops, G. C.; Samide, M. J. "Mass Spectrometry for the Masses," *J. Chem. Educ.* **2004**, *81*, 1169–1171.

Phillip, N.; Brennan, T.; Meleties; P. "Qualitative and Quantitative Anion Analysis of Drinking Water by Ion Chromatography," *Chem. Educ.* **2005**, *10* (3), 190–192.

Pidello, A. "Environmental Redox Potential and Redox Capacity Concepts Using a Simple Polarographic Experiment," *J. Chem. Educ.* **2003**, *80*, 68–70.

Quach, D. T.; Ciszkowski, N. A.; Finlayson-Pitts, B. J. "A New GC-MS Experiment for the Undergraduate Instrumental Analysis Laboratory in Environmental Chemistry: Methyl-*t*-butyl Ether and Benzene in Gasoline," *J. Chem. Educ.* **1998**, *75*, 1595–1598.

Radojevic, M.; Bashkin, V. N. *Practical Environmental Analysis*; The Royal Society of Chemistry: London, 1999.

Ramos, B. L.; Miller, S.; Korfmacher, K. "Implementation of a Geographic Information System in the Chemistry Curriculum: An Exercise in Integrating Environmental Analysis and Assessment," *J. Chem. Educ.* **2003**, *80*, 50–53.

Rice, J. K.; Jenkins, J. D.; Manley, C.; Sorel, E.; Smith, C. J. "Rapid Determination of Mercury in Seafood in an Introductory Environmental Science Class," *J. Chem. Educ.* **2005**, *82*, 265–268.

Rivera-Figueroa, A. M.; Ramazan, K. A.; Finlayson-Pitts, B. J. "Fluorescence, Absorption, and Excitation Spectra of Polycyclic Aromatic Hydrocarbons as a Tool for Quantitative Analysis," *J. Chem. Educ.* **2004**, *81*, 242–245.

Rudzinki, W. E.; Beu, S. "Gas Chromatographic Determination of Environmentally Significant Pesticides," *J. Chem. Educ.* **1982**, *59*, 614–615.

Rum, G.; Lee, W.-Y.; Gardea-Torresdey, J. "Applications of a U.S. EPA-Approved Method for Fluoride Determination in an Environmental Chemistry Laboratory: Fluoride Detection in Drinking Water," *J. Chem. Educ.* **2000**, *77*, 1604–1606.

Rump, H. H. *Laboratory Manual for the Examination of Water, Waste Water and Soil, 3rd ed.*; Wiley: New York, 2001.

Sadik, O. A.; Wanekaya, A. K.; Yevgeny, G. "Pressure-Assisted Chelating Extraction as a Teaching Tool in Instrumental Analysis," *J. Chem. Educ.* **2004**, *81*, 1177–1181.

Sawyer, C. N.; McCarty, P. L.; Parkin, G. F. *Chemistry for Environmental Engineering*; McGraw-Hill International Editions, Civil Engineering Series: New York, 1994.

• Gas Analysis (Chapter 32)

Schaumloffel, J. C. "Dendrochemical Analysis as a Teaching Tool in Analytical or Environmental Chemistry Courses," 214th. Am. Chem. Soc. National Meeting, Las Vegas, NV, Sept. 7–11, 1997. Chem. Ed. Division Paper # 002.

Shtoyko, T.; Zudans, I.; Seliskar, C. J.; Heineman, W. R.; Richardson, J. N. "An Attenuated Total Reflectance Sensor for Copper: An Experiment for Analytical or Physical Chemistry," *J. Chem. Educ.* **2004**, *81*, 1617–1619.

Salido, A.; Atterholt, C.; Bacon, J. R.; Butcher, D. J. "An Environmental Focus Using Inductively

Coupled Plasma Optical Emission Spectrometry and Ion Chromatography," *J. Chem. Educ.* **2003**, *80*, 22–23.

Seasholtz, M. B.; Pence, L. E.; Moe, O. A. "Determination of CO in Automobile Exhaust by FTIR Spectroscopy—An Instrumental Analysis Laboratory Experiment," *J. Chem. Educ.* **1988**, *65*, 820–823.

Seeley, J. V.; Bull, A. W.; Fehir, R. J., Jr.; Cornwall, S.; Knudsen, G. A.; Seeley, S. K. "A Simple Method for Measuring Ground-Level Ozone in the Atmosphere," *J. Chem. Educ.* **2005**, *82*, 282–285.

Seney, C. S.; Sinclair, K. V.; Bright, R. M.; Momoh, P. O.; Bozeman, A. D. "Development of a Multiple-Element Flame Emission Spectrometer Using CCD Detection," *J. Chem. Educ.* **2005**, *82*, 1826–1829.

Shane, E. C.; Price-Everett, M.; Hanson, T. "Fluorescence Measurement of Pyrene Wall Adsorption and Pyrene Association with Humic Acids. An Experiment for Physical Chemistry or Instrumental Methods," *J. Chem. Educ.* **2000**, *77*, 1617–1618.

Sinniah, K.; Piers, K. "Ion Chromatography: Analysis of Ions in Pond Waters," *J. Chem. Educ.* **2001**, *78*, 358–362.

Sittidech, M.; Street, S. "Environmental Analysis in the Instrumental Lab: More than One Way," *J. Chem. Educ.* **2003**, *80*, 376–377.

Snow, N. H.; Dunn, M.; Patel, S. "Determination of Crude Fat in Food Products by Supercritical Fluid Extraction and Gravimetric Analysis," *J. Chem. Educ.* **1997**, *74*, 1108–1111.

Solow, M. "Introduction of Mass Spectrometry in a First-Semester General Chemistry Laboratory Course: Quantification of MTBE or DMSO in Water," *J. Chem. Educ.* **2004**, *81*, 1172–1173.

Stemporzewski, S. E.; Butler, R. A.; Barry, E. F. "An Atomic Absorption Experiment for Environmental Chemistry," *J. Chem. Educ.* **1974**, *51*, 332–333.

Stephens, E. R.; Price, M. A. "Analysis of an Important Air Pollutant: Peroxylacetyl Nitrate (PAN)," *J. Chem. Educ.* **1973**, *50*, 351.

Subach, D. J.; Butwill, Bell, M. E. "Analytical Procedure for the Determination of Pesticides in Food," *J. Chem. Educ.* **1973**, *50*, 855–856.

Sundback, K. A. "Testing for Lead in the Environment," *J. Chem. Educ.* **1996**, *73*, 669.

Swinehart, J. H.; Mort, G. "Bringing Environmental Problems into the Science Classroom," *J. Coll. Sci. Teach.* **1995**, *24*, 58–61.

Todebush, P. M.; Geiger, F. M. "Sedimentation Time Measurements of Soil Particles by Light Scattering and Determination of Chromium, Lead, and Iron in Soil Samples via ICP," *J. Chem. Educ.* **2005**, *82*, 1542–1545.

University of Wisconsin-Stout, Chemistry Department. *Environmental Chemistry Lab Manual. Laboratory and Lecture Demonstrations.* http:/www.uwstout.edu/faculty/ondrusm/manual/index.html

Laboratory Experiments

- Copper and Arsenic in Treated Wood (Experiment 1)
- Effect of Heavy Metal Ions on the Growth of Microorganisms (Experiment 2)
- Analysis of Phosphate in Water (Experiment 3)
- Phosphates in Detergents (Experiment 4)
- Determination of Nitrate Ion in Water (Experiment 5)
- Ion Selective Electrodes (Experiment 6)
- Salts (Ionic Compounds) in Water (Experiment 7)
- Acidity and Alkalinity of Drinking Water (Experiment 8)
- Total Coliform Determination by Membrane Filtration (Experiment 9)
- Measurement of Dissolved Oxygen, BOD, and Rate of Oxygen Absorption in Water (Experiment 10)
- Identification of Food Dyes by Paper Chromatography (Experiment 11)
- Molecular Models (Experiment 12)
- Identification of FD&C Dyes by Visible Spectrophotometry (Experiment 13)
- Detection of Fuel Components by Gas Chromatography (Experiment 14)
- Determination of Heat of Combustion of Coal Using Bomb Calorimetry (Experiment 15)
- Heat of Combustion and Efficiency of Heat Transfer (Experiment 16)
- Measurement of Sulfur Content in Coal (Experiment 17)

- Detection of Polycyclic Aromatic Hydrocarbons in Water (Experiment 18)
- Oxides of Sulfur, Carbon, Phosphorus, Nitrogen, Magnesium, and Calcium (Experiment 19)
- Sampling of NO_x ($NO + NO_2$) and Particulates (Experiment 20)
- Preparation and Properties of Ozone (Experiment 21)
- Gas-Phase Analysis of Air Components and Air Pollutants Using FTIR (Experiment 22)
- Exponential Decay of a Transition Metal Complex Ion (Experiment 23)
- Single Digestion Procedure for Determining N, P, K, Ca, and Mg (Experiment 24)
  Laboratory and Lecture Demonstrations

- Catalytic Oxidation of Acetone (Demonstration 1)
- Methane Density and Flammability (Demonstration 2)
- Gases that are Denser than Air: Petroleum Ether and CO_2 (Demonstration 3)
- Tyndall Effect (Demonstration 4)
- Alum Treatment of Muddy Water (Demonstration 5)
- Electrostatic Precipitation of Smoke Particles (Demonstration 6)
- Surface Tension and Lowering it with a Surfactant (Demonstration 7)
- Bielstein Copper Wire Test for Halogenated Organics (Demonstration 8)
- Disappearing Polystyrene Popcorn (Demonstration 9)
- Combustion of Hydrocarbons and Alcohols (Demonstration 10)
- Explosive Behavior of Ethanol (Demonstration 11)

Urbanzky, E. T. "Carbinolamines and Geminal Diols in Aqueous Environmental Organic Chemistry," *J. Chem. Educ.* **2000**, *77*, 1644–1647.

Vlachogiannis, J. G.; Vlachonis, G. V. "An Experimental Design for the Determination of Cu and Pb in Marine Sediments Using Taguchi's Method," *Int. J. Environ. Anal. Chem.* **2003**, *83*, 1021–1034.

Volker, E. J.; DiLella, D.; Terneus, K.; Baldwin, C.; Volker, I. "The Determination of Ergosterol

in Environmental Samples. An Interdisciplinary Project Involving Techniques of Analytical and Organic Chemistry," *J. Chem. Educ.* **2000**, *77*, 1621–1623.

Vowles, P. D.; Connell, D.W. *Experiments in Environmental Chemistry: A Laboratory Manual* Pergamon Press: New York, 1980.

- Experiment 1. Photosynthesis, Respiration, and Biochemical Oxygen Demand
- Experiment 2. Eutrophication
- Experiment 3. Sewage Treatment—A Field Trip
- Experiment 4. Insecticides in Cigarette Smoke
- Experiment 5. Gas Chromatography of Volatile Hydrocarbons
- Experiment 6. Toxicity of Copper Ions toward Aquatic Biota
- Experiment 7. Lead in Household Paint
- Experiment 8. Atmospheric Pollutants
- Experiment 9. Aflatoxins in Peanuts
- Experiment 10. Food Additives
- Experiment 11. Erucic Acid Content of Bread
- Experiment 12. DDT in Human Milk
- Experiment 13. Chemical Defense of the Monarch Butterfly
- Experiment 14. Stream Pollution

Weidenhamer, J. D. "Environmental Projects in the Quantitative Analysis Lab," *J. Chem. Educ.* **1997**, *74*, 1437–1440.

Weinstein-Lloyd, J.; Lee, J. H. "Environmental Laboratory Exercise: Analysis of Hydrogen Peroxide by Fluorescence Spectroscopy," *J. Chem. Educ.* **1997**, *72*, 1053–1055.

Welch, L. E.; Mossman, D. M. "An Environmental Chemistry Experiment: The Determination of Radon Levels in Water," *J. Chem. Educ.* **1994**, *71*, 521–523.

Wenzel, T. J. "The Influence of Modern Instrumentation on the Analytical and General Chemistry Curriculum at Bates College," *J. Chem. Educ.* **2001**, *78*, 1164–1165.

Willey, J. D.; Avery, G. B., Jr.; Manock, J. J.; Skrabal, S. A.; Stehman, C. F. "Chemical Analysis of Soils: An Environmental Chemistry Laboratory for Undergraduate Science Majors," *J. Chem. Educ.* **1999**, *76*, 1693.

Wilson, R. I.; Mathers, D. T.; Mabury, S. A.; Jorgensen, G. M. "ELISA and GC- MS as

Teaching Tools in the Undergraduate Environmental Analytical Chemistry Laboratory," *J. Chem. Educ.* **2000**, *77*, 1619–1620.

Wingen, L. M.; Low, J. C.; Finlayson-Pitts, B. J. "Chromatography, Absorption, and Fluorescence: A New Instrumental Analysis Experiment on the Measurement of Polycyclic Aromatic Hydrocarbons in Cigarette Smoke," *J. Chem. Educ.* **1998**, *75*, 1599–1603.

Wong, J. W.; Ngim, K. K.; Eiserich, J. P.; Yeo, H. C. H.; Shibamoto, T.; Mabury, S. A. "Determination of Formaldehyde in Cigarette Smoke," *J. Chem. Educ.* **1997**, *74*, 1100–1003.

Woosley, R. S.; Butcher, D. J. "Chemical Analysis of an Endangered Conifer: Environmental Laboratory Experiments," *J. Chem. Educ.* **1998**, *75*, 1592–1594.

Xia, K.; Pierzynski, G. "Competitive Sorption between Oxalate and Phosphate in Soil: An Environmental Chemistry Laboratory Using Ion Chromatography," *J. Chem. Educ.* **2003**, *80*, 71–75.

Xiao, D.; Lin, L.; Yuan, H.; Choi, M. M.; Chan, W. "A Passive Sampler for Determination of Nitrogen Dioxide in Ambient Air," *J. Chem. Educ.* **2005**, *82*, 1231–1233.

Printed in the United States
By Bookmasters